10 South Carolina SC READY Grade 8 Math Practice Tests

The Ultimate Test Prep Collection with Answer Explanations

Dr. A. Nazari

10 Practice Tests

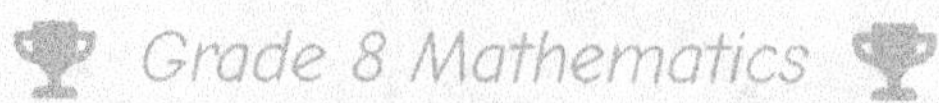

Welcome!

This book contains **10 full-length practice tests** — the most comprehensive preparation you can get for your Grade 8 math assessment. Each test covers all six topics:

📖 Irrational Numbers   📖 Powers & Scientific Notation

📖 Linear Equations   📖 Functions

📖 Geometry   📖 Data & Relationships

Ten tests give you the practice needed to walk into the real test feeling fully prepared.

Thorough preparation leads to outstanding results.

❝ Ten full tests! By the time you finish, there won't be any surprises on test day. **❞**

📖 How to Use This Book 📖

A complete 10-test preparation program

📋 What's Inside

- **10 Full-Length Practice Tests** — each covers all 6 chapters of Grade 8 math: irrational numbers, exponents & scientific notation, linear equations, functions, geometry, and data analysis.

- **Detailed Answer Explanations** — every question includes a step-by-step solution so you learn from every mistake.

- **Formula Reference Sheet** — all the key Grade 8 formulas you need, organized and ready for quick review.

- **Test Tracker** — log your scores across all 10 tests and monitor your progress from start to finish.

🕐 Your 10-Test Training Plan

⭐ PHASE 1: Foundation (Tests 1–3)

Untimed or soft-timed. Focus on understanding the format, identifying strengths and weaknesses, and building good study habits.

⭐⭐ PHASE 2: Building Skills (Tests 4–7)

Timed (70 minutes each). Work on pacing, accuracy, and showing complete solutions. Review weak topics between tests.

⭐⭐⭐ PHASE 3: Test-Day Ready (Tests 8–10)

Full test conditions: strict timing, quiet space, no notes. Compare scores with your early tests to see your growth.

Schedule: Take one test every 3–4 days, or one per week. Use study days between tests to review.

Types of Questions

 **Multiple Choice:** *Four options — work the problem first, then match. Eliminate obviously wrong answers to narrow your choices.*

Short Answer & Constructed Response: *Show every step: equations, substitutions, simplifications. Partial credit rewards correct reasoning even if the final answer is off.*

Graphing & Data Analysis: *Plot points, draw lines, interpret graphs. Label axes clearly.*

Tip: Ten tests is a full preparation program. Don't rush. The key is what you do between tests — study, review, and understand your mistakes before moving forward.

Find more at
ViewMath.com/SC-Grade8

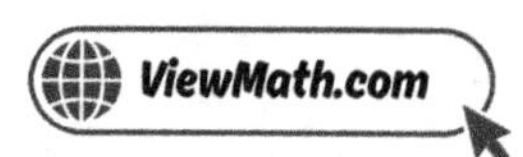

Test-Taking Tips

Before the Test

- Review your notes from the previous test — focus on your weak topics
- Set up a quiet, clean workspace with all your materials ready
- Start with a positive mindset: you've prepared for this

During the Test

- Read each problem fully before calculating anything
- Write the formula or set up the equation first, then substitute values
- Show all your work — every step, every operation
- If stuck for more than 2 minutes, mark it and move on
- Use estimation to check if your answers are reasonable

After the Test

- Read the full explanation for every question you got wrong
- Write down which topics gave you trouble (not just question numbers)
- Study those topics before taking the next test
- Record your score in the Test Tracker

Common Mistakes in Grade 8 Math

Exponents: $(ab)^n = a^n b^n$, but $a^m + a^n \neq a^{m+n}$. Only multiply/divide to combine.

Slope formula: $m = \frac{y_2 - y_1}{x_2 - x_1}$ — keep the order consistent.

Systems of equations: The solution must satisfy both equations.

Transformations: Rotations and reflections change position; dilations change size.

Volume: Use $\pi \approx 3.14$ or leave as π — match what the question asks.

The students who improve the most aren't the ones who take the most tests — they're the ones who carefully review every mistake. Make that your priority.

Find more at
ViewMath.com/SC-Grade8

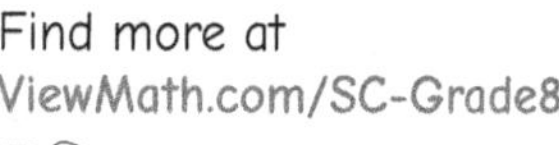

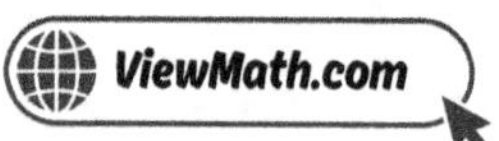

🧰 What You'll Need 🧰

Gather these materials before you begin

📋 Materials Checklist

Sharpened Pencils — #2 pencils, at least two

Good Eraser — for clean corrections

Scratch Paper — for working out problems

Ruler / Straightedge — for graphing & geometry

Quiet Space — no distractions

Focused Mind — ready to do your best

✓ Allowed Materials

✔ Pencils and eraser
✔ Scratch paper (provided on official test day)
✔ Ruler or straightedge (if required)
✔ Protractor (if required)

🚫 Not Allowed

✖ Calculator (unless your state test allows it)
✖ Cell phone or any electronic device
✖ Notes, textbooks, or reference sheets
✖ Help from others during the test

♥ A Note for Parents & Guardians

Ten tests is a comprehensive program. Plan **one test every 3–4 days** (or one per week) with study sessions between each test.

How to help:

- Tests 1–3 should be untimed — build understanding before adding pressure.
- After each test, review the answer explanations together. Ask: "Which topics were hardest? Let's study those before the next one."
- Use the Test Tracker to celebrate progress over time.
- For topic-specific help, pair this book with our **Grade 8 Math Study Guide** or **Grade 8 Workbook**.

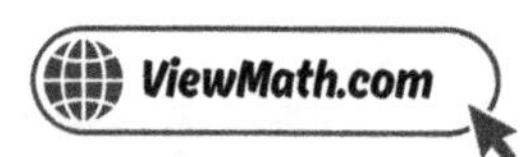

Grade 8 Formula Reference

X^1 Exponent Rules

$$a^m \cdot a^n = a^{m+n} \qquad (a^m)^n = a^{mn} \qquad (ab)^n = a^n \cdot b^n$$

$$\frac{a^m}{a^n} = a^{m-n} \qquad a^0 = 1 \ (a \neq 0) \qquad a^{-n} = \frac{1}{a^n}$$

Lines & Linear Equations

Slope: $m = \dfrac{y_2 - y_1}{x_2 - x_1} = \dfrac{rise}{run}$ **Slope-intercept:** $y = mx + b$ **Proportional:** $y = mx$

m = slope b = y-intercept   *Parallel lines: same slope*   *Proportional: passes through origin*

Scientific Notation

$a \times 10^n$ where $1 \leq |a| < 10$ **Multiply:** add exponents **Divide:** subtract exponents

$\sqrt{x}$ Roots & Number Sense

Perfect squares: 1, 4, 9, 16, 25, 36, 49, 64, 81, 100, 121, 144

Perfect cubes: 1, 8, 27, 64, 125 $\sqrt{2} \approx 1.414$ $\sqrt{3} \approx 1.732$ $\pi \approx 3.14159$

Pythagorean Theorem & Distance

$$a^2 + b^2 = c^2 \qquad c = \text{hypotenuse (longest side of a right triangle)}$$

Distance: $d = \sqrt{(x_2 - x_1)^2 + (y_2 - y_1)^2}$

Volume Formulas

Cylinder $V = \pi r^2 h$ **Cone** $V = \dfrac{1}{3}\pi r^2 h$ **Sphere** $V = \dfrac{4}{3}\pi r^3$

⬚ Angle Relationships

Triangle angle sum: $180°$ **Exterior angle** = sum of two remote interior angles

Parallel lines + transversal: Alternate interior angles are equal · Co-interior angles sum to $180°$

⬚ Functions

Each input $\rightarrow$ exactly one output   **Vertical line test:** if any vertical line hits graph more than once $\Rightarrow$ not a function

Linear: constant rate of change $(y = mx + b)$ **Nonlinear:** rate of change varies

⬚ Transformations

Translation: slide   **Reflection:** flip   **Rotation:** turn   **Dilation:** resize

Congruent = same shape & size   Similar = same shape, proportional size

 **Tip:** Bookmark this page! Review it before each test so these formulas become second nature.

Find more at
ViewMath.com/SC-Grade8

ViewMath.com

Grade 8 Formula Reference

Keep this page handy — you may use it during your practice tests!

X^1 Exponent Rules

$$a^m \cdot a^n = a^{m+n} \qquad (a^m)^n = a^{mn} \qquad (ab)^n = a^n \cdot b^n$$

$$\frac{a^m}{a^n} = a^{m-n} \qquad a^0 = 1 \ (a \neq 0) \qquad a^{-n} = \frac{1}{a^n}$$

Lines & Linear Equations

Slope: $m = \dfrac{y_2 - y_1}{x_2 - x_1} = \dfrac{rise}{run}$

m = slope $\quad$ b = y-intercept

Slope-intercept: $y = mx + b$

Parallel lines: same slope

Proportional: $y = mx$

Proportional: passes through origin

Scientific Notation

$a \times 10^n$ where $1 \leq |a| < 10$ $\qquad$ **Multiply:** add exponents $\qquad$ **Divide:** subtract exponents

$\sqrt{x}$ Roots & Number Sense

Perfect squares: 1, 4, 9, 16, 25, 36, 49, 64, 81, 100, 121, 144

Perfect cubes: 1, 8, 27, 64, 125 $\qquad \sqrt{2} \approx 1.414 \qquad \sqrt{3} \approx 1.732 \qquad \pi \approx 3.14159$

Pythagorean Theorem & Distance

$$a^2 + b^2 = c^2 \qquad c = \text{hypotenuse (longest side of a right triangle)}$$

Distance: $d = \sqrt{(x_2 - x_1)^2 + (y_2 - y_1)^2}$

Volume Formulas

Cylinder $V = \pi r^2 h \qquad$ **Cone** $V = \dfrac{1}{3}\pi r^2 h \qquad$ **Sphere** $V = \dfrac{4}{3}\pi r^3$

⬚ Angle Relationships

Triangle angle sum: 180°   **Exterior angle** = sum of two remote interior angles

Parallel lines + transversal: Alternate interior angles are equal · Co-interior angles sum to 180°

⬚ Functions

Each input → exactly one output   **Vertical line test:** if any vertical line hits graph more than once ⇒ not a function

Linear: constant rate of change $(y = mx + b)$ **Nonlinear:** rate of change varies

⟳ Transformations

Translation: slide   **Reflection:** flip   **Rotation:** turn   **Dilation:** resize

Congruent = same shape & size   *Similar = same shape, proportional size*

Tip: Bookmark this page! Review it before each test so these formulas become second nature.

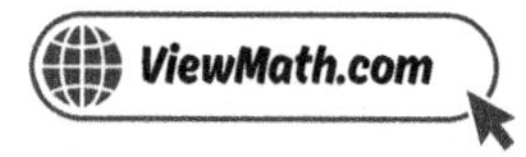

▦ Multiplication Table ▦

You may use this table during your practice tests!

×	1	2	3	4	5	6	7	8	9	10	11	12
1	1	2	3	4	5	6	7	8	9	10	11	12
2	2	4	6	8	10	12	14	16	18	20	22	24
3	3	6	9	12	15	18	21	24	27	30	33	36
4	4	8	12	16	20	24	28	32	36	40	44	48
5	5	10	15	20	25	30	35	40	45	50	55	60
6	6	12	18	24	30	36	42	48	54	60	66	72
7	7	14	21	28	35	42	49	56	63	70	77	84
8	8	16	24	32	40	48	56	64	72	80	88	96
9	9	18	27	36	45	54	63	72	81	90	99	108
10	10	20	30	40	50	60	70	80	90	100	110	120
11	11	22	33	44	55	66	77	88	99	110	121	132
12	12	24	36	48	60	72	84	96	108	120	132	144

💡 How to Use This Table

To find **4 × 7**:

1. Find **4** in the left column (blue).
2. Find **7** in the top row (blue).
3. Follow the row and column until they meet: the answer is **28**!

> ⓘ **Tip:** *You can also use this table for division! If you know* $28 \div 4 =$ *?, find 28 in the 4's row. The column header gives you the answer.* **7**!

Find more at
ViewMath.com/SC-Grade8

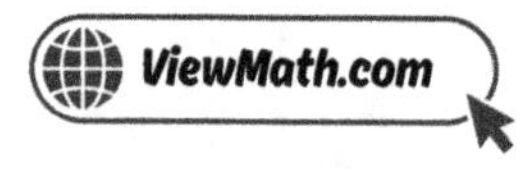

📈 My Test Tracker 📈

Record every test and watch your scores improve

Name: _______________________________   **Start Date:** _________________

GETTING STARTED (Tests 1–3)

Date: _____________   Score: _______/_______   %: _______   Topics to review: _____________________

Date: _____________   Score: _______/_______   %: _______   Topics to review: _____________________

Date: _____________   Score: _______/_______   %: _______   Topics to review: _____________________

BUILDING SKILLS (Tests 4–7)

Date: _____________   Score: _______/_______   %: _______   Focus area: _____________________

Date: _____________   Score: _______/_______   %: _______   Focus area: _____________________

Date: _____________   Score: _______/_______   %: _______   Focus area: _____________________

Date: _____________   Score: _______/_______   %: _______   Focus area: _____________________

TEST-DAY READY (Tests 8–10)

Date: _____________   Score: _______ / _______   %: _______   Growth since Test 1: _____________________

Date: _____________   Score: _______ / _______   %: _______   Growth since Test 1: _____________________

Date: _____________   Score: _______ / _______   %: _______   Growth since Test 1: _____________________

Score Progress

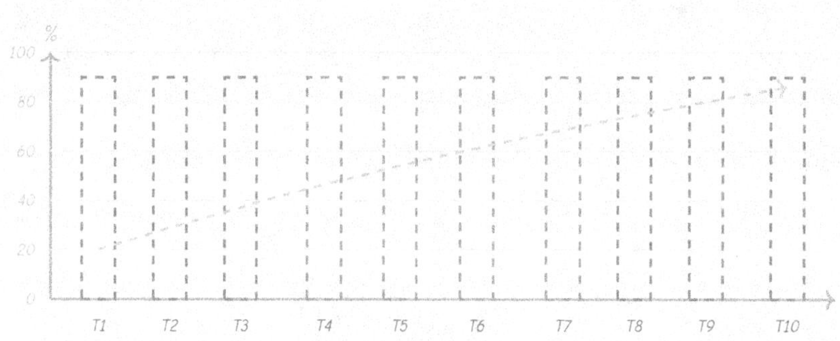

Shade each bar after every test. Watch your improvement!

Final Reflection

The most important thing I learned: ____________________________________

The topic where I improved the most: ____________________________________

My advice for other students: ____________________________________

Table of Contents

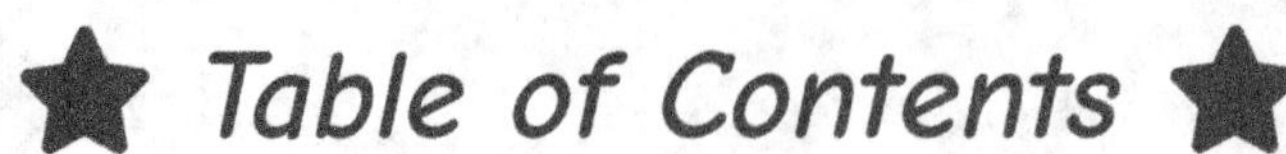

Here's what we'll explore together!

Let's learn and have fun!

Practice Test 1

📋 30 Questions

✏️ Before You Start ✏️

- ✓ **Read each question carefully** before choosing your answer.
- ✓ **Show your work** on scratch paper when you need to.
- ✓ **Skip hard questions** and come back to them later.
- ✓ **Check your answers** when you're done.
- ✓ **Take your time** — there's no rush!

⭐ You've Got This! ⭐

Do your best and show what you know!

1. True or false: The number $\frac{22}{7}$ is equal to π.

Your Answer

2. The table below shows four repeating decimals and the power of 10 a student used to convert each. Which one uses the WRONG power of 10?

Decimal	Multiply by
$0.\overline{7}$	10
$0.\overline{45}$	100
$0.\overline{123}$	100
$0.\overline{8}$	10

(A) $0.\overline{7}$

(B) $0.\overline{45}$

(C) $0.\overline{123}$

(D) $0.\overline{8}$

3. The table below shows a student's work to approximate $\sqrt{14}$. What should go in the blank?

Guess	Square	Compare to 14
3	9	too small
4	16	too big
3.7	13.69	too small
3.8	14.44	too big
?	?	closest

(A) 3.74, because $3.74^2 = 13.9876$

(B) 3.72, because $3.72^2 = 13.8384$

(C) 3.80, because $3.8^2 = 14.44$

(D) 3.75, because $3.75^2 = 14.0625$

Find more at
ViewMath.com/SC-Grade8

ViewMath.com

4. *What is the best estimate for $5 - \sqrt{7}$?*

(A) 1.4

(B) 2.4

(C) 3.0

(D) 0.4

5. *The diagram shows a square and a cube. The square has an area of A square units and the cube has a volume of V cubic units. If $A = V$, which pair of side lengths is correct?*

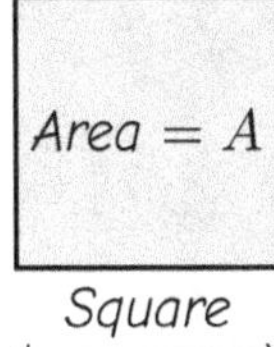

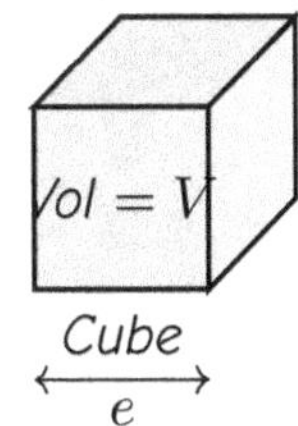

(A) $s = 6, e = 6$

(B) $s = 8, e = 4$

(C) $s = 9, e = 3$

(D) $s = 5, e = 5$

6. *Estimate the product $498{,}000{,}000 \times 6{,}200$ by first writing each factor in scientific notation and then computing.*

Your Answer:

Find more at
ViewMath.com/SC-Grade8

ViewMath.com

7. Study the diagram below. Each box produces its output by multiplying the two inputs. What belongs in the output box marked "?"?

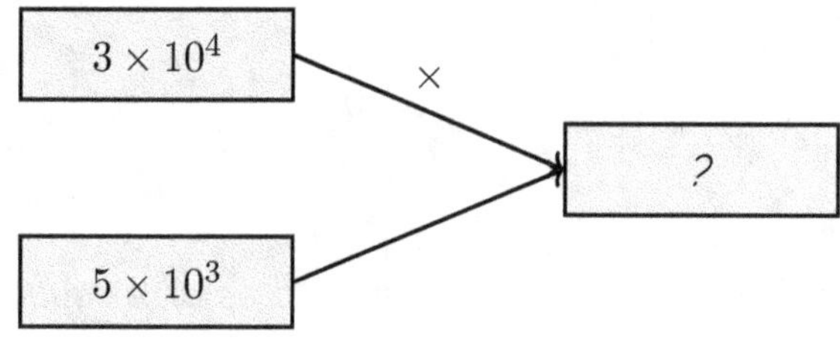

(A) 8×10^7

(B) 1.5×10^7

(C) 1.5×10^8

(D) 15×10^7

8. A bike rental company charges \$8 per hour. Which equation models the total cost y for x hours?

(A) $y = x + 8$

(B) $y = 8x$

(C) $y = \frac{x}{8}$

(D) $y = 8x + 10$

9. Find the slope through $(3, -1)$ and $(7, 11)$.

Your Answer:

10. A school play sold 200 tickets. Adult tickets were \$8 and student tickets were \$5. Total revenue was \$1,240. How many adult tickets were sold?

Your Answer:

11. *The graph below shows a relation. Is it a function?*

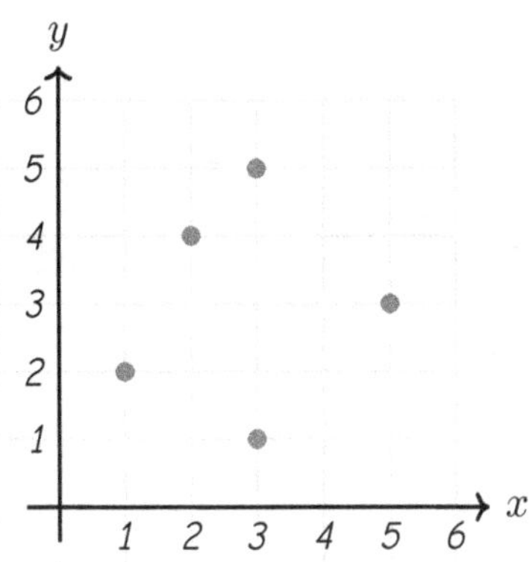

(A) *Yes, because all the points are in the first quadrant.*

(B) *Yes, because no two points have the same y-value.*

(C) *No, because there are two points at $x = 3$.*

(D) *No, because the points do not form a line.*

12. *The graph of h is shown below. For what value of x does $h(x) = 6$?*

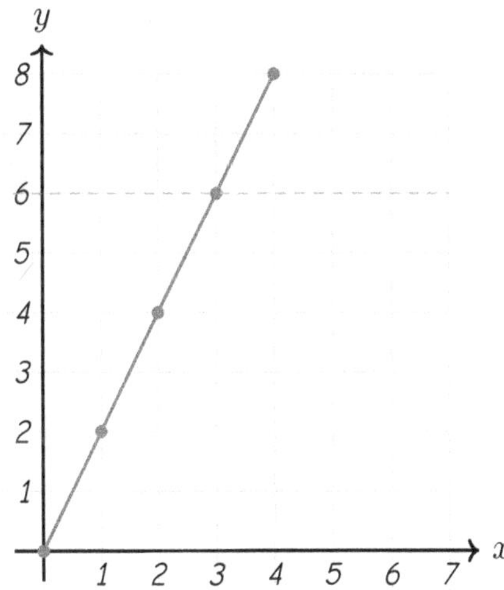

Your Answer:

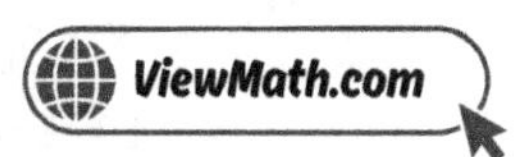

13. Function M has slope 4 and y-intercept 10. Function N has slope 4 and y-intercept 3. Are their graphs parallel? Write Yes or No.

Your Answer:

14. Which equation represents a linear function?

(A) $y = x^2 + 3$

(B) $y = \frac{1}{x}$

(C) $y = 4x - 7$

(D) $y = 2^x$

15. The table shows $(2, 9)$ and $(6, 25)$. Find the slope of the line.

Your Answer:

16. A student walks to a friend's house, waits for the friend, then they walk together to school. What does the distance-from-home graph look like?

(A) Increasing, constant, increasing

(B) Increasing, decreasing, increasing

(C) Constant, increasing, constant

(D) Decreasing, constant, decreasing

17. Name the three rigid transformations.

Your Answer:

18. $\triangle PQR \cong \triangle STU$. If $\angle P = 72°$ and $\angle Q = 53°$, what is $\angle U$?

Your Answer:

Find more at
ViewMath.com/SC-Grade8

ViewMath.com

19. A triangle with vertices $(0,0)$, $(6,0)$, and $(6,8)$ is dilated by factor $\frac{1}{2}$ from the origin. What is the perimeter of the image?

(A) 24

(B) 12

(C) 6

(D) 48

20. A dilation centered at the origin maps $(4,-2)$ to $(12,-6)$. What is the scale factor?

(A) 2

(B) 3

(C) 4

(D) 6

21. Which angles are equal when parallel lines are cut by a transversal?

(A) Co-interior angles

(B) Supplementary angles

(C) Corresponding angles

(D) Adjacent angles

22. A right triangle has legs 5 and 5. What is the hypotenuse?

(A) 10

(B) $\sqrt{50}$

(C) 25

(D) $\sqrt{10}$

23. What is the distance between $(-2,1)$ and $(4,9)$?

(A) 10

(B) $\sqrt{10}$

(C) 14

(D) 8

Find more at
ViewMath.com/SC-Grade8

ViewMath.com

24. A sphere's volume depends only on its:

(A) Height

(B) Diameter and height

(C) Radius

(D) Base area and height

25. As temperature increases, heating costs decrease. This relationship is:

(A) a positive linear association

(B) a negative linear association

(C) no association

(D) a nonlinear association

26. A scatter plot shows these points with a trend line through $(1, 3)$ and $(5, 7)$. What is the equation of the trend line?

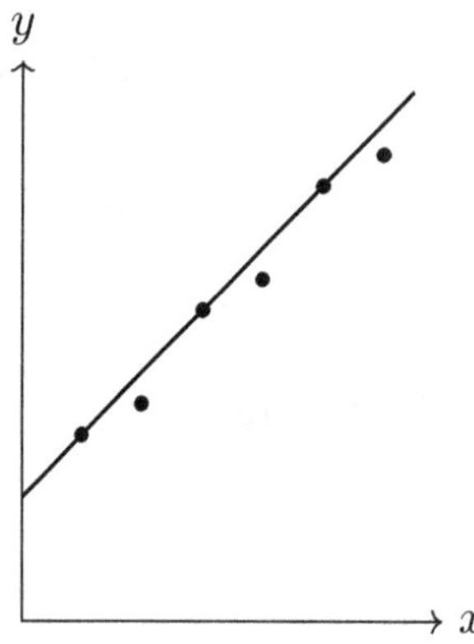

(A) $y = x + 2$

(B) $y = 2x + 1$

(C) $y = x + 3$

(D) $y = 0.5x + 2.5$

27. Why can extrapolation be unreliable?

(A) Because you cannot use a calculator

(B) Because the pattern may not continue outside the observed data range

(C) Because the slope is always wrong

(D) Because the y-intercept is always zero

28. Using the table from q19, what percentage of boys prefer reading? What percentage of girls prefer reading? Is there an association?

Your Answer

29. A MAD of 0 tells you that:

(A) The data values are all different

(B) All data values are the same

(C) The mean is 0

(D) There is one data point

30. A deck has 52 cards. Two cards are drawn without replacement. Find $P(\text{both aces})$.

Your Answer

Find more at
ViewMath.com/SC-Grade8

 # End of Practice Test 1 

Great job finishing the test!

 **My Score**

I got _____________ out of 30 questions right.

*Check your answers in the **Answer Key** at the back of the book.*

 *Review any questions you missed. That's how we learn!*

Check Your Score Online!

Visit **ViewMath Academy** to enter your answers and see which topics you need to review. You can also explore lessons, take quizzes, track your scores, and save your progress!

viewmath.com/score/8.1.SC.16

*Or go to **viewmath.com/score** and enter code: 8.1.SC.16*

2

Practice Test 2

📋 30 Questions

✏️ Before You Start ✏️

- ✓ **Read each question carefully** before choosing your answer.
- ✓ **Show your work** on scratch paper when you need to.
- ✓ **Skip hard questions** and come back to them later.
- ✓ **Check your answers** when you're done.
- ✓ **Take your time** — there's no rush!

⭐ You've Got This! ⭐

Do your best and show what you know!

1. The number line below shows two points, P and Q.

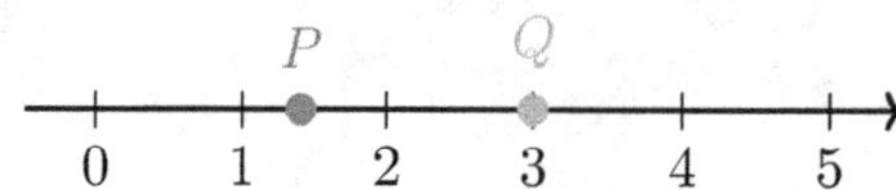

Point P represents $\sqrt{2}$ and point Q represents 3. Which statement is true?

(A) Both P and Q are rational.

(B) Both P and Q are irrational.

(C) P is irrational and Q is rational.

(D) P is rational and Q is irrational.

2. The flowchart below shows the conversion steps for a repeating decimal. Fill in the missing values to convert $0.\overline{54}$ to a fraction.

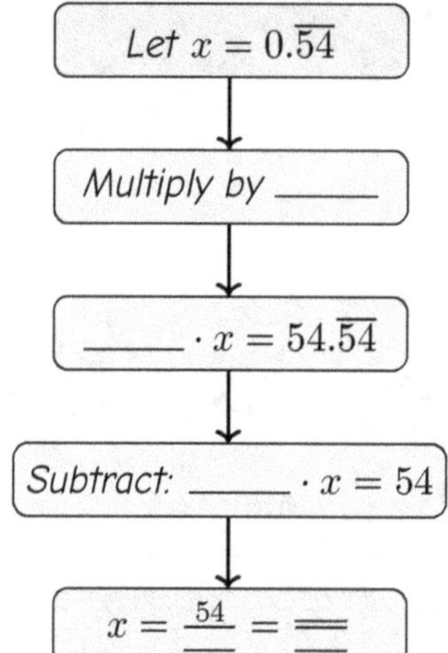

Your Answer:

3. The squares below have the given areas. Estimate the side length of each square to one decimal place.

A	B	C
Area $= 18$ sq units	Area $= 40$ sq units	Area $= 72$ sq units

Your Answer:

4. What is the best estimate for $\sqrt{5} + \sqrt{3}$?

(A) $\sqrt{8} \approx 2.8$

(B) 3.0

(C) 4.0

(D) 8.0

5. Solve $x^2 = \frac{9}{16}$.

(A) $x = \frac{3}{4}$ only

(B) $x = \pm\frac{3}{4}$

(C) $x = \frac{9}{8}$

(D) $x = \pm\frac{9}{4}$

6. Which of the following is true about 4.2×10^{-3} and 4.2×10^3?

(A) They are equal

(B) 4.2×10^3 is 10^6 times as large as 4.2×10^{-3}

(C) 4.2×10^{-3} is 10^6 times as large as 4.2×10^3

(D) Their product is 1

7. What is $\frac{8 \times 10^7}{2 \times 10^3}$?

(A) 4×10^4

(B) 6×10^4

(C) 4×10^{10}

(D) 16×10^{10}

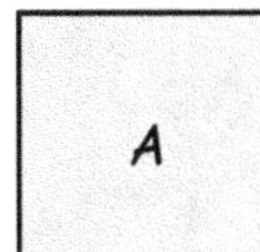

Find more at
ViewMath.com/SC-Grade8

ViewMath.com

8. A faucet leaks at a constant rate. It leaks 45 mL in 15 minutes. How many mL does it leak in 1 hour?

Your Answer:

9. A plant grows from 4 cm to 16 cm in 6 weeks. What is the rate of change?

(A) 1 cm/week

(B) 2 cm/week

(C) 3 cm/week

(D) 4 cm/week

10. You have $5 bills and $10 bills totaling $85. You have 12 bills. How many $10 bills do you have?

Your Answer:

11. Given $\{(-2, 5), (-1, 3), (0, 1), (1, -1)\}$, what is the output when the input is 0?

(A) -1

(B) 0

(C) 1

(D) 5

12. A function table shows $f(1) = 4$, $f(2) = 7$, $f(3) = 10$. What is $f(2) - f(1)$?

Your Answer:

Find more at
ViewMath.com/SC-Grade8

13. The tables below show two functions. Find the rate of change and initial value of each. Which function will have a greater value at $x = 10$?

Function A

x	y
0	20
1	23
2	26
3	29

Function B

x	y
0	5
1	10
2	15
3	20

Your Answer:

14. Give an example of a nonlinear equation.

Your Answer:

15. A line passes through $(1, 6)$ and $(4, 18)$. What is the slope?

(A) 3

(B) 4

(C) 6

(D) 12

16. A bathtub fills up, then someone soaks, then the water drains. Which describes the water level graph?

(A) Decreasing, constant, increasing

(B) Increasing, decreasing, constant

(C) Increasing, constant, decreasing

(D) Constant, increasing, decreasing

17. A figure is reflected over the y-axis. What happens to point $(7, -2)$?

(A) $(7, 2)$

(B) $(-7, 2)$

(C) $(-7, -2)$

(D) $(-2, 7)$

Find more at
ViewMath.com/SC-Grade8

ViewMath.com

18. Triangle $ABC \cong$ Triangle DEF. If $AB = 8$ cm, what is DE?

   (A) 4 cm   (B) 8 cm

   (C) 16 cm   (D) Cannot be determined

19. Translate the point $(4, -6)$ by $(-3, 8)$. What is the image?

   Your Answer:

20. A rectangle is 3 cm by 5 cm. Another rectangle is 6 cm by 9 cm. Are they similar?

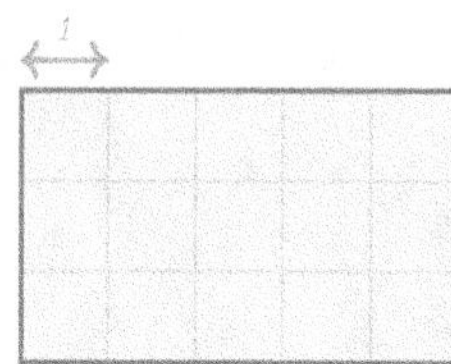

   (A) Yes, with scale factor 2   (B) Yes, with scale factor 3

   (C) No, the sides are not proportional.   (D) Yes, with scale factor $\frac{3}{5}$

21. Two angles of a triangle are $37°$ and $53°$. What is the exterior angle adjacent to the third angle?

   (A) $90°$ (B) $127°$

   (C) $53°$ (D) $143°$

22. Is a triangle with sides 5, 10, 12 a right triangle?

   Your Answer:

Find more at
ViewMath.com/SC-Grade8

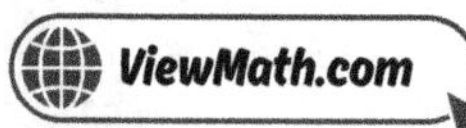

23. *What is the distance between* $(0,0)$ *and* $(5,0)$*?*

(A) 0

(B) 5

(C) 25

(D) $\sqrt{5}$

24. *A cone has radius 5 cm and height 12 cm. What is its volume in terms of* π*?*

Your Answer:

25. *Identify the outlier in the scatter plot.*

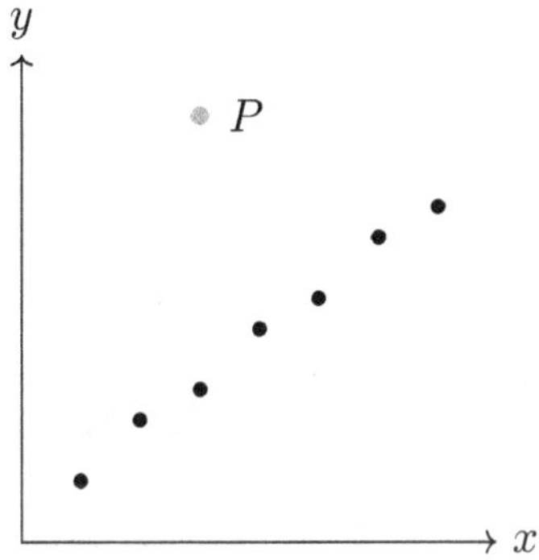

(A) $(1,1)$

(B) $(7,5.5)$

(C) $(3,7)$ — *point* P

(D) $(4,3.5)$

26. Draw a line of best fit through the data and estimate its slope.

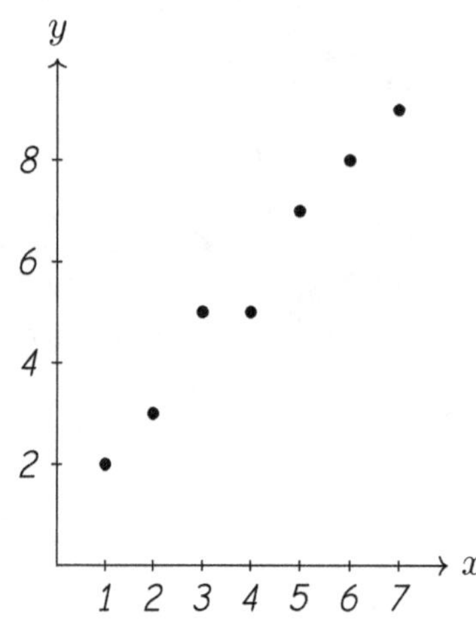

Your Answer:

27. A linear model is $y = 2x + 5$. What does the slope represent?

(A) The starting value

(B) The rate of change per unit increase in x

(C) The x-intercept

(D) The total value of y

28. A two-way table has 3 rows and 4 columns (not counting totals). How many joint frequencies are there?

(A) 3

(B) 4

(C) 7

(D) 12

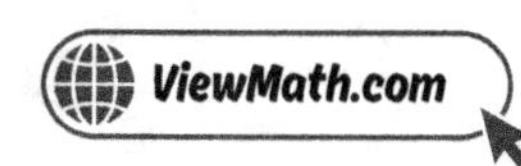

29. *A number line shows data points $\{3, 5, 5, 7, 10\}$ with the mean marked at 6. What is the MAD?*

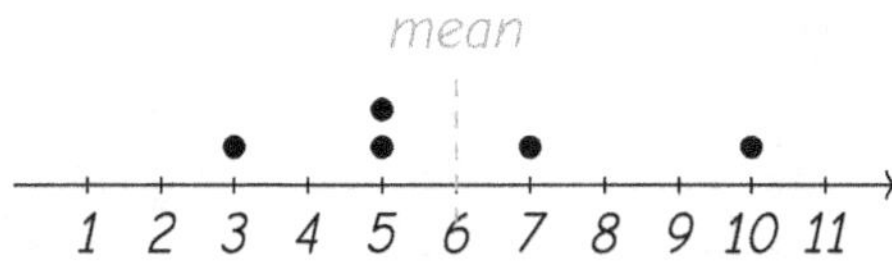

(A) 1.6

(B) 2

(C) 2.2

(D) 3

30. *A tree diagram shows flipping a coin and rolling a die (1–6). How many total outcomes are in the sample space?*

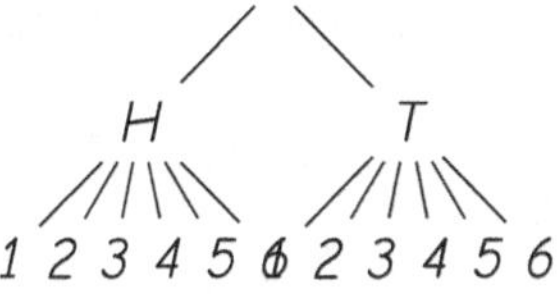

(A) 6

(B) 8

(C) 12

(D) 36

 # End of Practice Test 2 

Great job finishing the test!

 My Score

I got _____________ out of 30 questions right.

*Check your answers in the **Answer Key** at the back of the book.*

Review any questions you missed. That's how we learn!

 Check Your Score Online!

Visit **ViewMath Academy** to enter your answers and see which topics you need to review. You can also explore lessons, take quizzes, track your scores, and save your progress!

viewmath.com/score/8.1.SC.17

Or go to viewmath.com/score and enter code: 8.1.SC.17

Practice Test 3

30 Questions

✏ Before You Start ✏

- ✓ **Read each question carefully** before choosing your answer.
- ✓ **Show your work** on scratch paper when you need to.
- ✓ **Skip hard questions** and come back to them later.
- ✓ **Check your answers** when you're done.
- ✓ **Take your time** — there's no rush!

⭐ You've Got This! ⭐

Do your best and show what you know!

1. Which number below can be written as a fraction $\frac{a}{b}$ where a and b are integers and $b \neq 0$?

   (A) $\sqrt{3}$

   (B) π

   (C) $0.\overline{81}$

   (D) $\sqrt{11}$

2. What is $0.\overline{27}$ as a fraction in simplest form?

   (A) $\frac{27}{100}$

   (B) $\frac{27}{99}$

   (C) $\frac{3}{11}$

   (D) $\frac{9}{33}$

3. A student estimated $\sqrt{60} \approx 8$. Is this estimate reasonable?

   (A) Yes, because $8^2 = 64$ is close to 60.

   (B) No, because $8^2 = 64$, which is too far from 60. A better estimate is about 7.7.

   (C) No, $\sqrt{60} = 30$.

   (D) Yes, because $60 \div 8 \approx 8$.

4. Which is larger: $4\sqrt{2}$ or $\sqrt{30}$?

   (A) $4\sqrt{2}$, because $4 > \sqrt{30}$

   (B) $\sqrt{30}$, because $30 > 2$

   (C) $4\sqrt{2}$, because $4\sqrt{2} \approx 5.66 > 5.48 \approx \sqrt{30}$

   (D) They are equal.

5. A square has an area of 196 square centimeters. What is the length of one side?

   (A) 13 cm

   (B) 49 cm

   (C) 14 cm

   (D) 98 cm

6. Which of the following is 0.00072 written in scientific notation?

   (A) 7.2×10^{-4}

   (B) 7.2×10^{4}

   (C) 72×10^{-5}

   (D) 0.72×10^{-3}

Find more at
ViewMath.com/SC-Grade8

ViewMath.com

7. A virus has a mass of 9.5×10^{-18} grams. If a sample contains 2×10^6 viruses, what is the total mass?

(A) 1.9×10^{-11} g

(B) 1.9×10^{-12} g

(C) 11.5×10^{-12} g

(D) 1.9×10^{-24} g

8. Store A sells apples for $2 per pound. Store B's prices are shown in the table.

Pounds	3	6	9
Cost ($)	7.50	15.00	22.50

Which store has the lower unit price?

(A) Store A

(B) Store B

(C) They charge the same price.

(D) Not enough information.

9. A car's odometer reads 120 miles at 2:00 PM and 270 miles at 5:00 PM. What is the car's speed (rate of change)?

(A) 45 mph

(B) 50 mph

(C) 60 mph

(D) 90 mph

10. Two trains leave the same station heading in opposite directions. Train A goes 60 mph and Train B goes 80 mph. After how many hours are they 420 miles apart?

(A) 2

(B) 2.5

(C) 3

(D) 3.5

11. In the ordered pair (x, y), x is called the:

(A) output

(B) range

(C) input

(D) function

Find more at
ViewMath.com/SC-Grade8

ViewMath.com

12. If $f(x) = 4x + 1$, find the value of $f(3) + f(1)$.

Your Answer:

13. Function A starts at \$50 and increases by \$12 per week. Function B: $y = 15x + 30$. Which earns more per week?

(A) Function A

(B) Function B

(C) They earn the same per week.

(D) Cannot be determined.

14. Is $y = x^3 - 2$ linear or nonlinear?

Your Answer:

15. A phone plan costs \$35 per month with no activation fee. Which equation models the total cost?

(A) $y = 35$

(B) $y = 35x$

(C) $y = 35x + 35$

(D) $y = x + 35$

16. A roller coaster goes up, down sharply, up again, then levels off. How many sections does the graph have?

(A) 2

(B) 3

(C) 4

(D) 5

Find more at
ViewMath.com/SC-Grade8

17. The figure below shows a point P and its image P' after a transformation. Which transformation was applied?

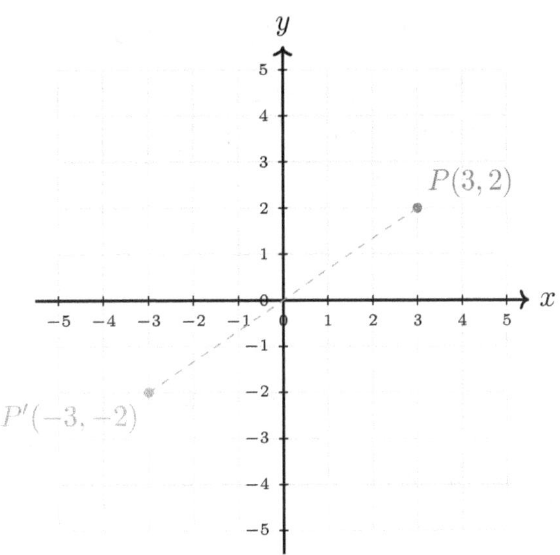

(A) Reflection over the x-axis

(B) Reflection over the y-axis

(C) Rotation 90° counterclockwise

(D) Rotation 180° around the origin

18. Which transformation would NOT preserve congruence?

(A) Reflection over the x-axis

(B) Translation 2 units up

(C) Dilation by a factor of 1.5

(D) Rotation of 270°

19. A point $(-1, 6)$ is rotated 90° clockwise around the origin. The 90° clockwise rule is $(x, y) \rightarrow (y, -x)$. What is the image?

(A) $(6, 1)$

(B) $(-6, -1)$

(C) $(-6, 1)$

(D) $(1, -6)$

20. *Are all squares similar to each other?*

(A) *No, because they can have different side lengths.*

(B) *No, because they can have different angles.*

(C) *Yes, because all squares have equal angles and proportional sides.*

(D) *Yes, but only if they have the same perimeter.*

21. *An exterior angle of a triangle is always:*

(A) *Less than any interior angle*

(B) *Equal to the adjacent interior angle*

(C) *Greater than either non-adjacent interior angle*

(D) *Equal to $180°$*

22. *Find the missing side x in the right triangle below.*

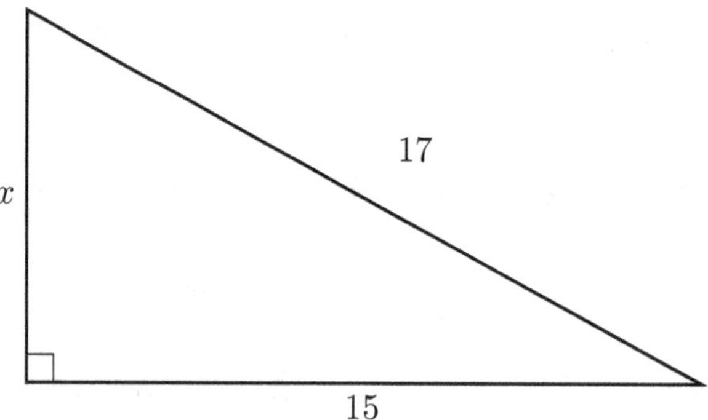

Your Answer:

23. *Two corners of a square are at $(0,0)$ and $(0,6)$. What is the length of the diagonal?*

(A) 6

(B) 12

(C) $6\sqrt{2}$

(D) 36

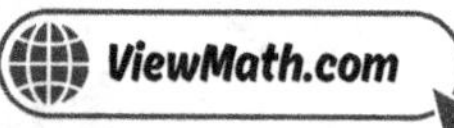

24. A basketball has a diameter of 24 cm. What is its volume? Use $\pi \approx 3.14$.

(A) $7{,}234.6\ cm^3$

(B) $2{,}143.6\ cm^3$

(C) $904.3\ cm^3$

(D) $18{,}086.4\ cm^3$

25. A scatter plot has a strong upward trend. Which r-value (correlation) is most likely?

(A) $r = -0.9$

(B) $r = 0.1$

(C) $r = 0.9$

(D) $r = 0$

26. Data: $(1, 2), (2, 4), (3, 7), (4, 8), (5, 11)$. Estimate the slope of a reasonable trend line.

Your Answer:

27. A positive residual means:

(A) The predicted value is higher than the actual value.

(B) The actual value is higher than the predicted value.

(C) The slope of the model is positive.

(D) The data point is an outlier.

28. A survey asks students about their favorite sport (soccer or basketball) and grade (7th or 8th). 20 7th-graders chose soccer. This 20 is a:

(A) marginal frequency

(B) joint frequency

(C) relative frequency

(D) total frequency

29. Data: $\{2, 4, 6, 8, 10\}$. The mean is 6. What are the absolute deviations?

(A) $4, 2, 0, 2, 4$

(B) $-4, -2, 0, 2, 4$

(C) $2, 4, 6, 8, 10$

(D) $6, 6, 6, 6, 6$

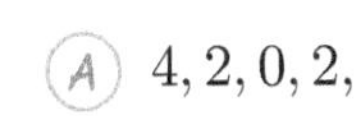

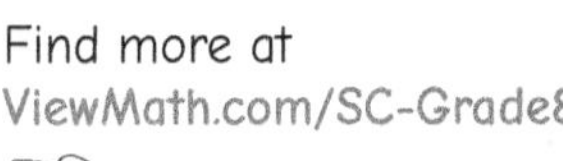
Find more at
ViewMath.com/SC-Grade8

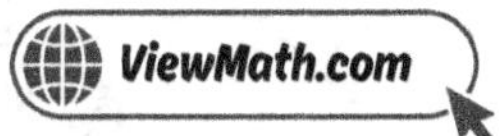

30. *A fair die is rolled twice. What is $P(\text{both rolls are } 6)$?*

(A) $\frac{1}{6}$

(B) $\frac{1}{12}$

(C) $\frac{1}{36}$

(D) $\frac{2}{6}$

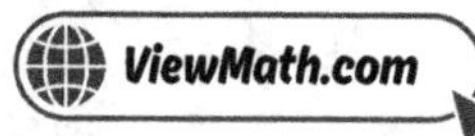

 *End of Practice Test 3* 

Great job finishing the test!

 My Score

I got _____________ out of 30 questions right.

*Check your answers in the **Answer Key** at the back of the book.*

💡 Review any questions you missed. That's how we learn!

📊 Check Your Score Online!

Visit **ViewMath Academy** to enter your answers and see which topics you need to review. You can also explore lessons, take quizzes, track your scores, and save your progress!

viewmath.com/score/8.1.SC.18

Or go to viewmath.com/score and enter code: 8.1.SC.18

Practice Test 4

✎ 30 Questions

✏ Before You Start ✏

- ✓ **Read each question carefully** before choosing your answer.
- ✓ **Show your work** on scratch paper when you need to.
- ✓ **Skip hard questions** and come back to them later.
- ✓ **Check your answers** when you're done.
- ✓ **Take your time** — there's no rush!

★ You've Got This! ★

Do your best and show what you know!

1. Which of the following numbers is irrational?

(A) $\frac{5}{8}$ (B) $0.\overline{6}$

(C) $\sqrt{9}$ (D) $\sqrt{7}$

2. Write $0.5\overline{8}$ as a fraction in simplest form.

Your Answer

3. Which inequality is true?

(A) $\sqrt{15} < 3$ (B) $\sqrt{15} > 4$

(C) $\sqrt{15} > 3$ (D) $\sqrt{15} = 4$

4. Compute $\sqrt{2} \times \sqrt{18}$ exactly, without a calculator.

Your Answer

5. What is $\sqrt{49}$?

(A) 6 (B) 7

(C) 8 (D) 24.5

6. What is 6.1×10^3 written in standard form?

(A) 61,000 (B) 610

(C) 6,100 (D) 0.0061

7. A bacterium has a length of 2×10^{-6} m. If 1 micrometer (μm) $= 10^{-6}$ m, what is the bacterium's length in micrometers?

(A) $2 \times 10^{-12}\ \mu m$

(B) $0.002\ \mu m$

(C) $2\ \mu m$

(D) $2{,}000\ \mu m$

8. Taxi A charges \$3 per mile. Taxi B charges \$36 for a 9-mile trip. Which taxi has the higher unit rate?

Your Answer:

9. What is the slope through $(-4, -1)$ and $(2, 5)$?

(A) -1

(B) $\frac{2}{3}$

(C) 1

(D) $\frac{3}{2}$

10. Two numbers add to 25 and differ by 7. What is the larger number?

(A) 14

(B) 15

(C) 16

(D) 17

11. Does the equation $y = -4x + 9$ represent a function?

(A) No, because the slope is negative.

(B) No, because y can be negative.

(C) Yes, because each value of x gives exactly one value of y.

(D) Yes, because the equation has two variables.

12. If $f(x) = 8x - 6$ and $f(x) = 42$, what is x?

Your Answer:

Find more at
ViewMath.com/SC-Grade8

13. The table shows Function P. Function Q is $y = 2x + 9$. Which function has the greater initial value?

x	0	1	2	3
$P(x)$	5	8	11	14

(A) Function P

(B) Function Q

(C) They have the same initial value.

(D) Cannot be determined.

14. A table shows $(1, 1)$, $(2, 8)$, $(3, 27)$, $(4, 64)$. Is this linear or nonlinear? What pattern do you see?

Your Answer

15. The function $y = 50x + 200$ models a bank account balance after x weeks. What is the weekly deposit?

(A) $50

(B) $100

(C) $200

(D) $250

16. Which graph best matches this story: "Maria walks quickly to the store, waits in line, then walks slowly home"?

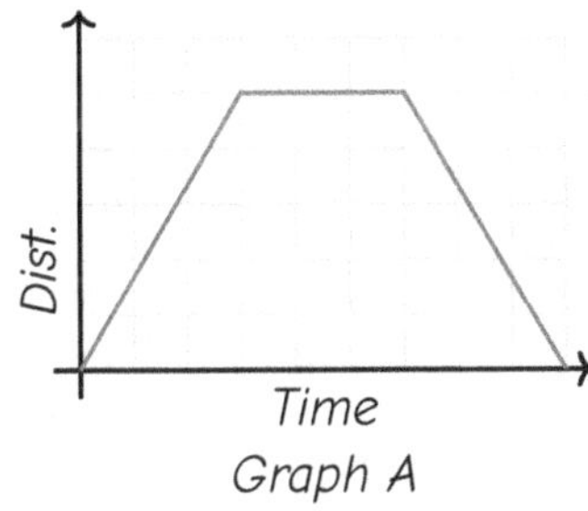

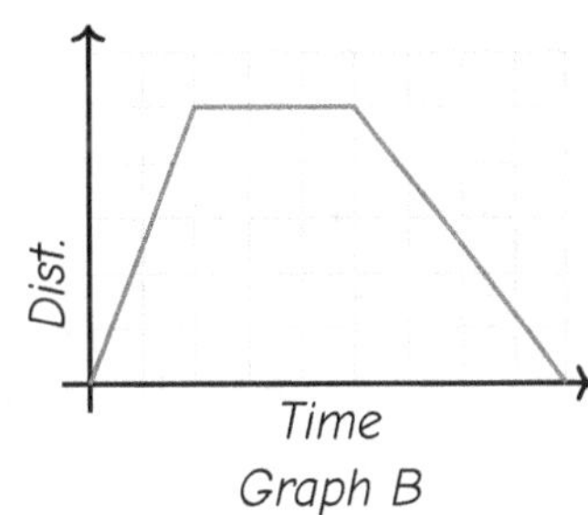

(A) Graph A only

(B) Graph B only

(C) Both graphs match equally.

(D) Neither graph matches.

Find more at
ViewMath.com/SC-Grade8

17. Which transformation turns a figure around a fixed point?

(A) Translation

(B) Reflection

(C) Rotation

(D) Dilation

18. A student says two circles with radius 5 cm are always congruent. Is this correct?

(A) No, circles cannot be congruent.

(B) No, they need the same center too.

(C) Yes, because they have the same radius.

(D) Yes, but only if they are in the same position.

19. Point (a, b) is rotated $180°$ and then reflected over the x-axis. The final image is:

(A) (a, b)

(B) $(-a, b)$

(C) $(a, -b)$

(D) $(-a, -b)$

20. A dilation with $k = 1$ produces:

(A) A figure twice as large

(B) A congruent figure

(C) A figure half as large

(D) No figure at all

21. Two parallel lines are cut by a transversal. One angle measures $72°$. What is its alternate interior angle?

(A) $18°$

(B) $72°$

(C) $108°$

(D) $288°$

Find more at
ViewMath.com/SC-Grade8

22. A ladder 15 ft long leans against a wall. The base is 9 ft from the wall. How high up the wall does the ladder reach?

(A) 6 ft

(B) 12 ft

(C) $\sqrt{306}$ ft

(D) 24 ft

23. Points P and Q are shown on the grid. What is the distance between them?

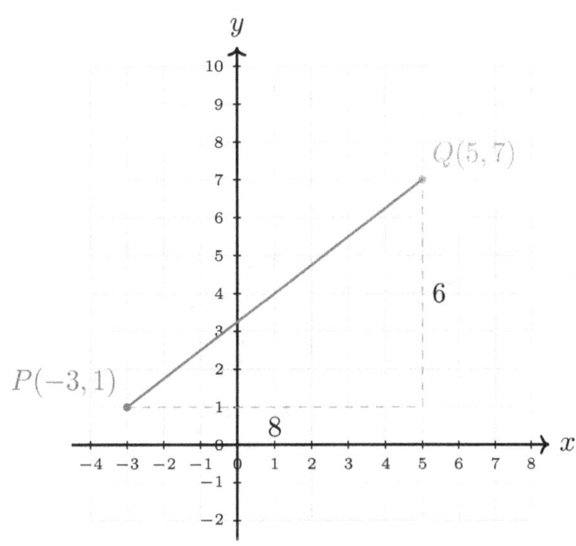

(A) 14

(B) $\sqrt{14}$

(C) 10

(D) $\sqrt{28}$

24. A cone has volume 100 cm^3. A cylinder with the same base and height has volume ___ cm^3.

Your Answer:

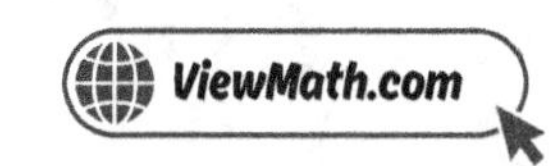

25. A scatter plot of shoe size versus favorite color shows dots scattered randomly with no pattern. What type of association is this?

  (A) Positive

  (B) Negative

  (C) Linear

  (D) No association

26. A trend line passes through $(3, 7)$ and $(9, 19)$. What is the slope?

  (A) 1

  (B) 2

  (C) 3

  (D) 4

27. In the model $y = 2.5x + 1$, the actual value at $x = 4$ is $y = 12$. What is the residual?

  (A) 0

  (B) 1

  (C) -1

  (D) 11

28.

	Pass	Fail	Total
Studied	30	5	35
Did not study	10	15	25
Total	40	20	60

What fraction of all students passed?

  (A) $\frac{30}{60}$

  (B) $\frac{40}{60}$

  (C) $\frac{35}{60}$

  (D) $\frac{20}{60}$

Find more at
ViewMath.com/SC-Grade8

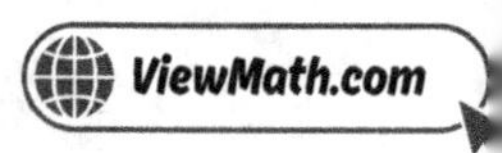

29. *Two data sets are shown on dot plots. Set X: $\{5, 5, 5, 5, 5\}$. Set Y: $\{1, 3, 5, 7, 9\}$. Calculate the MAD for each set and explain the comparison.*

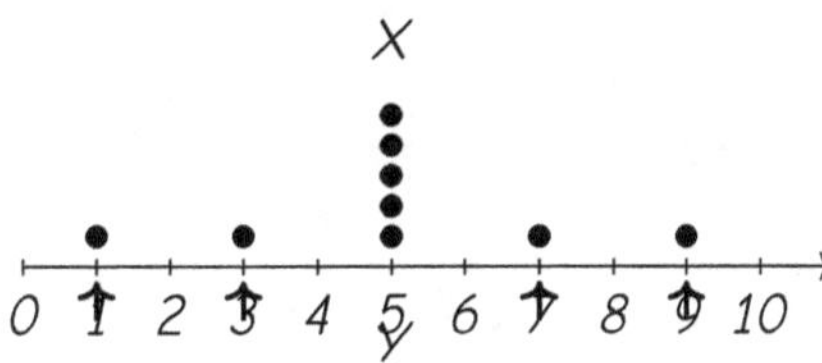

Your Answer

30. *A jar has 6 green and 4 yellow balls. One ball is drawn and replaced. Then another is drawn.* $P(both\ green) =?$

(A) $\frac{36}{100}$

(B) $\frac{30}{90}$

(C) $\frac{6}{10}$

(D) $\frac{12}{20}$

 # ⭐ *End of Practice Test 4* ⭐ 

Great job finishing the test!

 My Score

I got _____________ out of 30 questions right.

*Check your answers in the **Answer Key** at the back of the book.*

💡 Review any questions you missed. That's how we learn!

📊 Check Your Score Online!

Visit **ViewMath Academy** to enter your answers and see which topics you need to review. You can also explore lessons, take quizzes, track your scores, and save your progress!

viewmath.com/score/8.1.SC.19

Or go to viewmath.com/score and enter code: 8.1.SC.19

5

Practice Test 5

📋 30 Questions

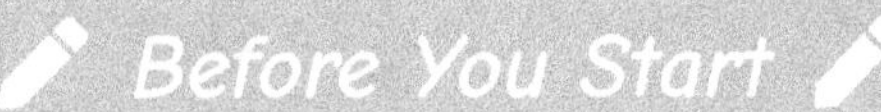

✏️ Before You Start ✏️

- ✔ **Read each question carefully** before choosing your answer.
- ✔ **Show your work** on scratch paper when you need to.
- ✔ **Skip hard questions** and come back to them later.
- ✔ **Check your answers** when you're done.
- ✔ **Take your time** — there's no rush!

 You've Got This! 

Do your best and show what you know!

1. A calculator shows $\sqrt{144} = 12$. Is $\sqrt{144}$ rational or irrational? Explain.

Your Answer:

2. To convert $0.\overline{63}$ to a fraction, which equation should you set up after letting $x = 0.\overline{63}$?

(A) $10x - x = 63$

(B) $100x - x = 63$

(C) $100x - x = 6.3$

(D) $1000x - x = 63$

3. Which inequality correctly describes the location of $\sqrt{22}$?

(A) $4 < \sqrt{22} < 5$

(B) $5 < \sqrt{22} < 6$

(C) $10 < \sqrt{22} < 12$

(D) $\sqrt{22} = 11$

4. The diagram below shows a rectangle with irrational side lengths. Which is the best estimate of the area?

$\sqrt{20}$ cm

$\sqrt{8}$ cm

(A) $\sqrt{28} \approx 5.3$ cm^2

(B) $\sqrt{160} \approx 12.6$ cm^2

(C) 28 cm^2

(D) 160 cm^2

5. Which of the following is a perfect square?

(A) 50

(B) 72

(C) 81

(D) 90

Find more at
ViewMath.com/SC-Grade8

6. A hydrogen atom has a radius of about 2.5×10^{-11} meters. How many places do you move the decimal to write this in standard form?

Your Answer:

7. What is $5.2 \times 10^6 + 3.8 \times 10^6$?

(A) 9×10^6

(B) 9×10^{12}

(C) 19.76×10^6

(D) 9×10^{36}

8. Which ordered pair could NOT lie on the graph of a proportional relationship?

(A) $(0, 0)$

(B) $(1, 5)$

(C) $(3, 15)$

(D) $(2, 13)$

9. Find the slope through $\left(\frac{1}{2}, 3\right)$ and $\left(\frac{3}{2}, 7\right)$.

(A) 2

(B) 4

(C) $\frac{1}{4}$

(D) 8

10. A rectangle has a perimeter of 48 m. Its length is twice its width. Find the dimensions.

Your Answer:

11. Is the relation $\{(3, 8), (4, 8), (5, 8), (6, 8)\}$ a function? Write Yes or No.

Your Answer:

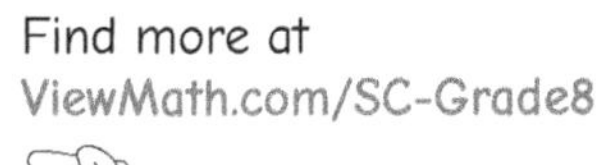

Find more at
ViewMath.com/SC-Grade8

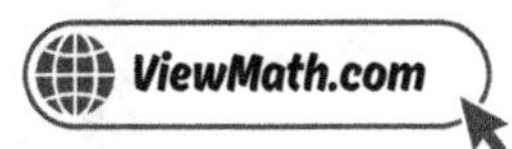

12. If $f(x) = -2x + 10$, what is $f(0)$?

(A) -2

(B) 0

(C) 8

(D) 10

13. A table for Function C shows $(0, 7)$, $(1, 11)$, $(2, 15)$. What is the initial value?

Your Answer:

14. A student says $y = 3x^2$ is linear because the coefficient is 3. What is wrong with this reasoning?

(A) The coefficient should be 1 for it to be linear.

(B) The variable x is raised to the 2nd power, making it nonlinear.

(C) The function has no y-intercept.

(D) Nothing is wrong; it is linear.

15. A line passes through $(2, 10)$ and $(5, 22)$. What is the equation?

(A) $y = 4x + 2$

(B) $y = 3x + 4$

(C) $y = 4x + 10$

(D) $y = 12x - 2$

16. A graph is flat for 4 seconds on a speed-time graph. What does this mean?

Your Answer:

17. A rectangle has vertices at $(1, 2)$, $(5, 2)$, $(5, 4)$, and $(1, 4)$. After a translation of 3 units right and 1 unit up, what are the coordinates of the vertex that was at $(1, 2)$?

(A) $(4, 3)$

(B) $(4, 1)$

(C) $(2, 5)$

(D) $(-2, 3)$

Find more at
ViewMath.com/SC-Grade8

ViewMath.com

18. *Figure B is obtained by reflecting Figure A over a line and then translating it 3 units right. Are the figures congruent?*

(A) No, because two transformations were used.

(B) No, because the position changed.

(C) Yes, because both transformations are rigid.

(D) Yes, but only if the figures are triangles.

19. *Which transformation maps (x, y) to $(-y, x)$?*

(A) Reflection over the x-axis

(B) Reflection over the y-axis

(C) $90°$ counterclockwise rotation

(D) $180°$ rotation

20. *If two figures are congruent, are they also similar?*

(A) Yes, with scale factor $k = 1$.

(B) No, congruent and similar are different properties.

(C) Only if they are triangles.

(D) Only if they have the same orientation.

21. *In a triangle, the angles are $(2x + 5)°$, $(3x)°$, and $(x + 25)°$. Find x.*

Your Answer:

22. *Is a triangle with sides 7, 24, 25 a right triangle?*

(A) Yes, because $7 + 24 > 25$.

(B) Yes, because $7^2 + 24^2 = 25^2$.

(C) No, because $7^2 + 24^2 \neq 25^2$.

(D) No, because all sides must be equal.

23. *A line segment has endpoints $(2, 3)$ and $(10, 9)$. What is its length?*

(A) 10

(B) $\sqrt{10}$

(C) 14

(D) 8

Find more at
ViewMath.com/SC-Grade8

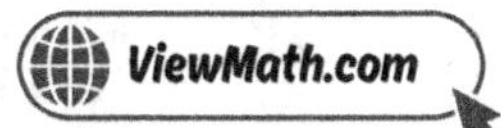

24. What is the volume of a cylinder with radius 4 cm and height 10 cm? Use $\pi \approx 3.14$.

(A) $125.6\ cm^3$ (B) $502.4\ cm^3$

(C) $160\ cm^3$ (D) $251.2\ cm^3$

25. Which scatter plot pattern suggests a linear association?

(A) Dots that form a U-shape   (B) Dots scattered randomly

(C) Dots that roughly follow a straight line   (D) Dots that form a circle

26. Which describes a "good fit"?

(A) All data points are far from the line.   (B) Data points are close to the line.

(C) All data points are below the line.   (D) The line is vertical.

27. A trend line gives $y = -3x + 45$. Data was collected for $x = 1$ to $x = 10$. Predict y when $x = 5$ and when $x = 30$. State which prediction is more reliable and why.

Your Answer

28. A relative frequency table shows the value 0.15 in one cell. If the total is 200, what is the frequency?

(A) 15   (B) 30

(C) 10   (D) 20

Find more at
ViewMath.com/SC-Grade8

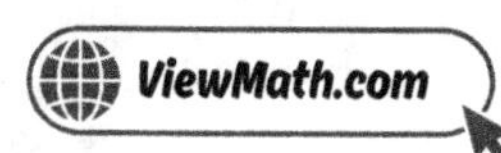

29. *Data:* $\{1, 3, 5, 7, 9\}$. *Plot the data on the number line below, find the mean, mark it, and compute the MAD.*

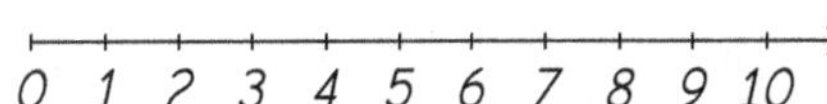

Your Answer:

30. $P(A) = 0.6$, $P(B) = 0.4$. *If A and B are independent, $P(A$ and $B) = ?$*

(A) 1.0

(B) 0.24

(C) 0.20

(D) 0.10

 # End of Practice Test 5 

Great job finishing the test!

 My Score

I got _____________ out of 30 questions right.

*Check your answers in the **Answer Key** at the back of the book.*

💡 *Review any questions you missed. That's how we learn!*

📊 Check Your Score Online!

Visit **ViewMath Academy** to enter your answers and see which topics you need to review. You can also explore lessons, take quizzes, track your scores, and save your progress!

viewmath.com/score/8.1.SC.20

*Or go to **viewmath.com/score** and enter code: 8.1.SC.20*

Practice Test 6

30 Questions

✏ Before You Start ✏

- ✓ **Read each question carefully** before choosing your answer.
- ✓ **Show your work** on scratch paper when you need to.
- ✓ **Skip hard questions** and come back to them later.
- ✓ **Check your answers** when you're done.
- ✓ **Take your time** — there's no rush!

★ You've Got This! ★

Do your best and show what you know!

1. Is $\frac{5}{6}$ rational or irrational? Explain.

Your Answer:

2. When converting $0.\overline{123}$ to a fraction, by what power of 10 should you multiply both sides?

(A) 10

(B) 100

(C) 1000

(D) 10000

3. Between which two consecutive integers does $\sqrt{110}$ lie?

(A) 9 and 10

(B) 10 and 11

(C) 11 and 12

(D) 54 and 56

4. A student says $\sqrt{4} + \sqrt{9} = \sqrt{13}$. Is this correct?

(A) Yes, because you add the numbers under the square roots.

(B) No, $\sqrt{4} + \sqrt{9} = 2 + 3 = 5$, but $\sqrt{13} \approx 3.6$.

(C) Yes, because $4 + 9 = 13$.

(D) No, $\sqrt{4} + \sqrt{9} = \sqrt{36} = 6$.

5. A cube has a volume of 343 cubic inches. What is the length of one edge?

(A) 7 in.

(B) 49 in.

(C) $114.\overline{3}$ in.

(D) 17 in.

6. What is 9.03×10^{-5} written in standard form?

(A) 903,000

(B) 0.0000903

(C) 0.000903

(D) 0.00903

Find more at
ViewMath.com/SC-Grade8

7. A light year is approximately 9.5×10^{12} km. If a star is 4 light years away, how far is it in kilometers?

(A) 3.8×10^{13} km

(B) 3.8×10^{12} km

(C) 13.5×10^{12} km

(D) 9.5×10^{48} km

8. A graph shows a straight line through $(0,0)$ and $(2,9)$. What is the unit rate?

(A) 2

(B) 4.5

(C) 9

(D) 18

9. A line passes through $(2,7)$ and $(5,1)$. What is the slope?

(A) -2

(B) 2

(C) -3

(D) 3

10. Maria has 15 coins, all nickels and dimes. The coins are worth \$1.10. How many dimes does she have?

(A) 5

(B) 6

(C) 7

(D) 8

11. Which of these equations represents a function?

(A) $x = 5$

(B) $y = 7$

(C) $x^2 + y^2 = 16$

(D) $x = y^2$

12. Use the graph of f below to find $f(2)$.

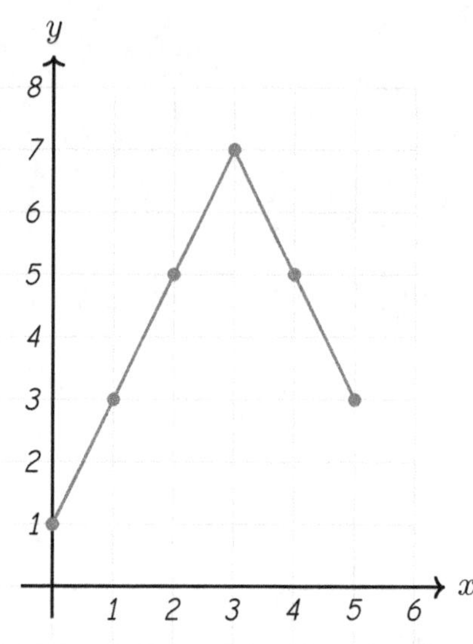

(A) 2

(B) 3

(C) 5

(D) 7

13. Function A: $y = -x + 10$. Function B: $y = -4x + 10$. Which function decreases faster? Write A or B.

Your Answer

14. Which table shows a linear function?

(A) $(1, 1), (2, 4), (3, 9), (4, 16)$

(B) $(0, 5), (1, 8), (2, 11), (3, 14)$

(C) $(1, 2), (2, 4), (3, 8), (4, 16)$

(D) $(0, 1), (1, 3), (2, 9), (3, 27)$

15. The table below shows a linear function. What is the equation?

x	0	1	2	3
y	-4	-1	2	5

(A) $y = 3x - 4$

(B) $y = -4x + 3$

(C) $y = -3x + 4$

(D) $y = 3x + 4$

16. Which graph could represent a person standing still for 5 seconds, then walking at a steady pace?

(A) A line sloping upward for the entire time

(B) A horizontal line, then a line sloping upward

(C) A line sloping downward, then a horizontal line

(D) A curve going upward the entire time

17. A triangle is translated 6 units left and 3 units up. A vertex was at $(2, -1)$. What are its new coordinates?

Your Answer:

18. True or false: If two figures have the same area, they must be congruent.

Your Answer:

19. Point $(3, 4)$ is dilated by a factor of 2 from the origin. What is the image?

(A) $(5, 6)$

(B) $(6, 8)$

(C) $(1.5, 2)$

(D) $(6, 4)$

20. Two similar pentagons have a scale factor of $\frac{2}{3}$. If the perimeter of the larger pentagon is 45 cm, what is the perimeter of the smaller?

Your Answer:

21. Corresponding angles formed by a transversal and two parallel lines are located:

(A) On opposite sides of the transversal, between the parallel lines

(B) On the same side of the transversal, in matching positions

(C) On opposite sides of the transversal, outside the parallel lines

(D) Adjacent to each other at one intersection

22. Is a triangle with sides 11, 60, 61 a right triangle?

Your Answer:

23. What is the distance between $(-5, 0)$ and $(7, 0)$?

(A) -12

(B) 2

(C) 12

(D) $\sqrt{74}$

24. What is the volume of a cylinder with radius 7 m and height 3 m? Leave your answer in terms of π.

(A) $21\pi \ m^3$

(B) $63\pi \ m^3$

(C) $147\pi \ m^3$

(D) $49\pi \ m^3$

25. Which variable typically goes on the x-axis in a scatter plot?

(A) The dependent variable

(B) The response variable

(C) The independent (explanatory) variable

(D) The variable with the largest values

Find more at
ViewMath.com/SC-Grade8

26. *A line passes through $(0,3)$ and $(4,11)$. What is the y-intercept?*

(A) 0

(B) 4

(C) 3

(D) 11

27. *Which statement about slope in a linear model is true?*

(A) *A positive slope means y decreases as x increases.*

(B) *A negative slope means y increases as x increases.*

(C) *A slope of zero means y stays constant.*

(D) *The slope is always positive in real data.*

28. *What is a relative frequency?*

(A) *The total number of data values*

(B) *A frequency expressed as a fraction or percentage of a total*

(C) *The difference between two frequencies*

(D) *The mode of the data*

29. *Data: $\{4, 4, 4, 4, 4\}$. What is the MAD?*

(A) 4

(B) 1

(C) 0

(D) 2

30. *What is the probability of an impossible event?*

(A) 1

(B) 0.5

(C) 0

(D) −1

Find more at
ViewMath.com/SC-Grade8

 End of Practice Test 6 

Great job finishing the test!

 My Score

I got __________ out of 30 questions right.

*Check your answers in the **Answer Key** at the back of the book.*

💡 *Review any questions you missed. That's how we learn!*

📊 Check Your Score Online!

Visit **ViewMath Academy** to enter your answers and see which topics you need to review. You can also explore lessons, take quizzes, track your scores, and save your progress!

viewmath.com/score/8.1.SC.21

Or go to viewmath.com/score and enter code: 8.1.SC.21

Practice Test 7

 30 Questions

✏️ Before You Start ✏️

✓ **Read each question carefully** before choosing your answer.

✓ **Show your work** on scratch paper when you need to.

✓ **Skip hard questions** and come back to them later.

✓ **Check your answers** when you're done.

✓ **Take your time** — there's no rush!

 ⭐ You've Got This! 

Do your best and show what you know!

1. Give an example of an irrational number between 1 and 2.

   Your Answer:

2. A student converts $0.\overline{36}$ and gets $\frac{36}{100}$. What mistake did the student make?

   (A) The student divided by 100 instead of 99.

   (B) The student forgot to simplify.

   (C) The student treated a repeating decimal as a terminating decimal.

   (D) The student multiplied by 10 instead of 100.

3. Between which two consecutive integers does $\sqrt{83}$ lie?

   Your Answer:

4. A student claims that $2\sqrt{5} = \sqrt{10}$. Is this correct?

   (A) Yes, because $2 \times 5 = 10$.

   (B) No, $2\sqrt{5} = \sqrt{20}$, not $\sqrt{10}$.

   (C) Yes, multiplication distributes into the square root.

   (D) No, $2\sqrt{5} = 4\sqrt{5}$.

5. A cube-shaped box has a volume of 216 cubic centimeters. What is the edge length of the box?

   Your Answer:

6. Which of the following is 350,000 written in scientific notation?

   (A) 35×10^4

   (B) 3.5×10^5

   (C) 3.5×10^4

   (D) 0.35×10^6

Find more at
ViewMath.com/SC-Grade8

ViewMath.com

7. Earth's mass is about 6×10^{24} kg and Jupiter's mass is about 1.9×10^{27} kg. About how many times more massive is Jupiter than Earth? Round to the nearest whole number.

Your Answer:

8. Which table does NOT represent a proportional relationship?

A)

x	1	2
y	4	8

B)

x	2	4
y	5	10

C)

x	3	6
y	9	15

D)

x	5	10
y	15	30

9. What is the slope of the line through $(1, 3)$ and $(4, 12)$?

A) 2

B) 3

C) 4

D) 9

10. A rectangle's perimeter is 34 cm. The length is 5 cm more than the width. What is the width?

A) 4 cm

B) 5 cm

C) 6 cm

D) 7 cm

11. How many times can a vertical line cross the graph of a function at most?

Your Answer:

12. A function is defined by the table below.

x	0	1	2	3
$f(x)$	−1	3	7	11

What is $f(0) + f(2)$?

(A) 2

(B) 6

(C) 8

(D) 10

13. Two functions have the same rate of change but different initial values. What is true about their graphs?

(A) They are the same line.

(B) They are parallel lines.

(C) They intersect at the origin.

(D) One is curved and the other is straight.

14. A table shows $(1, 3)$, $(2, 6)$, $(3, 11)$, $(4, 18)$. Is the function linear or nonlinear?

(A) Linear, because y increases as x increases.

(B) Linear, because the first difference is 3.

(C) Nonlinear, because the differences in y are $3, 5, 7$ — not constant.

(D) Nonlinear, because all outputs are positive.

15. Write the equation of a line with slope 6 and y-intercept -3.

Your Answer:

16. A flat (horizontal) section of a graph of height vs. time means:

(A) The height is increasing.

(B) The height is decreasing.

(C) The height is staying the same.

(D) Time has stopped.

Find more at
ViewMath.com/SC-Grade8

17. A figure is moved 4 units to the right and 2 units down without changing its size or shape. What transformation is this?

(A) Rotation

(B) Reflection

(C) Translation

(D) Dilation

18. $\triangle ABC \cong \triangle DEF$. If $\angle A = 55°$ and $\angle B = 80°$, what is $\angle F$?

(A) $55°$

(B) $80°$

(C) $45°$

(D) $135°$

19. A triangle has a vertex at $(8, -4)$. After a dilation by factor $\frac{1}{2}$ from the origin, what is the image of this vertex?

(A) $(16, -8)$

(B) $(4, -2)$

(C) $(4, -4)$

(D) $(8, -2)$

20. Triangle A has sides 6, 8, 10. Triangle B has sides 9, 12, 15. What is the scale factor from A to B?

Your Answer

21. In an isosceles triangle, the two base angles are each 54°. What is the vertex angle?

(A) $54°$

(B) $72°$

(C) $108°$

(D) $126°$

22. Is a triangle with sides 6, 8, 11 a right triangle?

(A) Yes

(B) No, because $6^2 + 8^2 \neq 11^2$

(C) No, because $6 + 8 \neq 11$

(D) Yes, because all sides are different lengths.

Find more at
ViewMath.com/SC-Grade8

23. *Find the distance between $(3, 1)$ and $(7, 4)$.*

Your Answer

24. *What is the volume of the cone shown below in terms of π?*

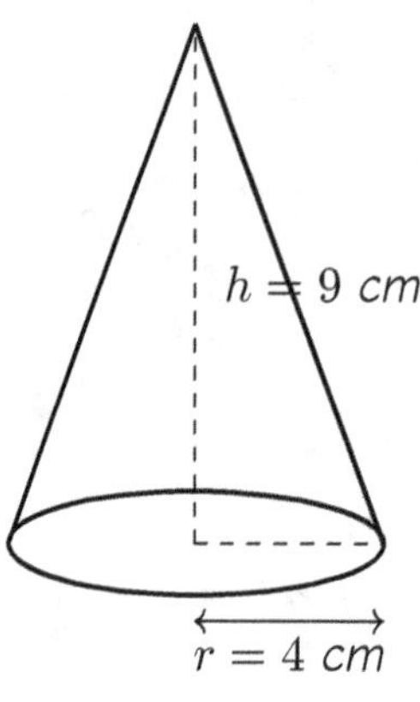

(A) 144π cm^3

(B) 48π cm^3

(C) 36π cm^3

(D) 108π cm^3

25. *Does a strong association in a scatter plot prove that one variable causes the other? Explain.*

Your Answer

26. *A scatter plot shows a strong nonlinear (curved) pattern. Should you draw a straight line of best fit?*

(A) *Yes — always draw a straight line.*

(B) *No — a straight line does not fit curved data well.*

(C) *Yes — but only if there are outliers.*

(D) *No — you should never draw any line.*

27. A model for plant height is $y = 0.4x + 2$ where x is days and y is height in cm. Using the graph, find the predicted height at day 10, and explain what the slope and y-intercept mean.

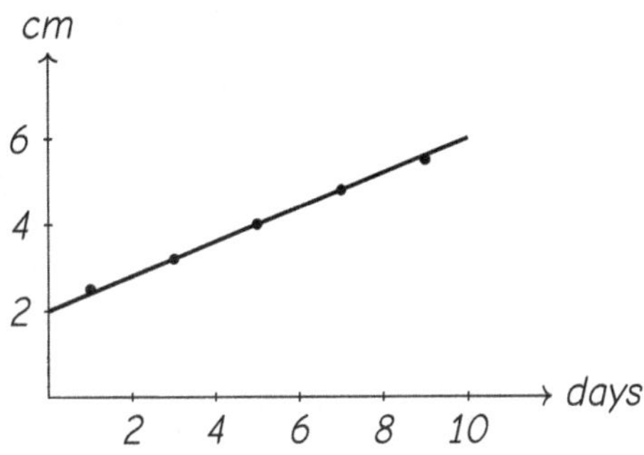

Your Answer:

28. What is a two-way frequency table?

(A) A table with two rows

(B) A table that shows frequencies for two categorical variables

(C) A table that shows only percentages

(D) A table with exactly two columns

29. Two classes took the same test. Class X had MAD = 5 and Class Y had MAD = 12. Which class had more consistent scores?

(A) Class X

(B) Class Y

(C) Both were equally consistent

(D) Cannot be determined

30. A card is drawn from a standard 52-card deck. What is $P(\text{heart})$?

(A) $\frac{1}{52}$

(B) $\frac{1}{13}$

(C) $\frac{1}{4}$

(D) $\frac{1}{2}$

Find more at
ViewMath.com/SC-Grade8

ViewMath.com

Great job finishing the test!

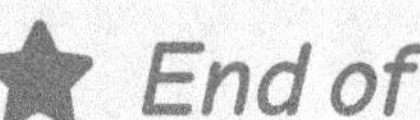 My Score

I got _____________ out of 30 questions right.

*Check your answers in the **Answer Key** at the back of the book.*

 *Review any questions you missed. That's how we learn!*

 Check Your Score Online!

Visit **ViewMath Academy** to enter your answers and see which topics you need to review. You can also explore lessons, take quizzes, track your scores, and save your progress!

viewmath.com/score/8.1.SC.22

Or go to viewmath.com/score and enter code: 8.1.SC.22

8

Practice Test 8

📋 *30 Questions*

✏️ Before You Start ✏️

- ✓ **Read each question carefully** before choosing your answer.
- ✓ **Show your work** on scratch paper when you need to.
- ✓ **Skip hard questions** and come back to them later.
- ✓ **Check your answers** when you're done.
- ✓ **Take your time** — there's no rush!

⭐ You've Got This! ⭐

Do your best and show what you know!

1. Which of the following is a true statement?

   (A) Every square root is irrational.   (B) Every integer is irrational.

   (C) Every integer is rational.   (D) Every decimal is irrational.

2. The table below shows repeating decimals and their fraction equivalents. Find the missing fraction for $0.\overline{81}$.

Decimal	Unsimplified Fraction	Simplest Form
$0.\overline{27}$	$\frac{27}{99}$	$\frac{3}{11}$
$0.\overline{54}$	$\frac{54}{99}$	$\frac{6}{11}$
$0.\overline{81}$	?	?

Your Answer:

3. Approximate $\sqrt{45}$ to one decimal place.

Your Answer:

4. Which expression equals exactly 6?

   (A) $\sqrt{3} \times \sqrt{12}$ (B) $\sqrt{3} + \sqrt{12}$

   (C) $\sqrt{6} \times \sqrt{6}$ (D) Both A and C

5. Evaluate $\sqrt[3]{-125}$.

Your Answer:

Find more at
ViewMath.com/SC-Grade8

ViewMath.com

6. The bar chart compares the masses of four animals. Which animal's mass is closest to 1×10^2 kilograms?

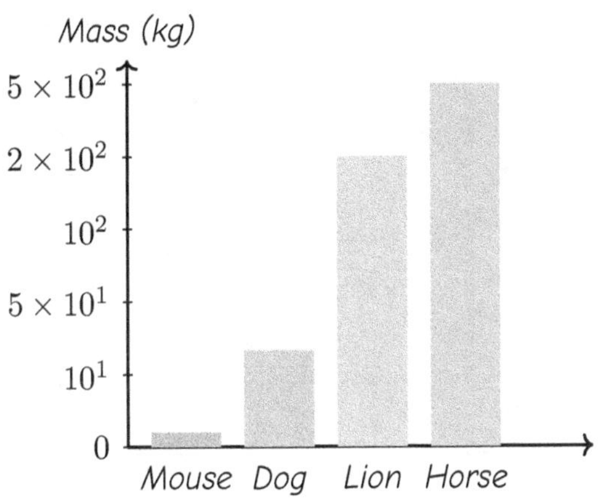

(A) Mouse

(B) Dog

(C) Lion

(D) Horse

7. A phone stores 6.4×10^{10} bytes. A photo uses 4×10^6 bytes. How many photos can the phone store?

(A) 1.6×10^3

(B) 1.6×10^4

(C) 1.6×10^{16}

(D) 2.4×10^4

8. Machine A fills $y = 12x$ bottles per hour. Machine B fills 50 bottles in 5 hours. Which machine is faster?

(A) Machine A

(B) Machine B

(C) They fill at the same rate.

(D) Cannot be determined.

9. The temperature drops from 68°F to 50°F over 6 hours. What is the rate of change in temperature?

(A) 3°F per hour

(B) −3°F per hour

(C) 6°F per hour

(D) −6°F per hour

10. One number is 4 more than another. Their sum is 36. Find both numbers.

Your Answer:

11. Which graph below represents a function?

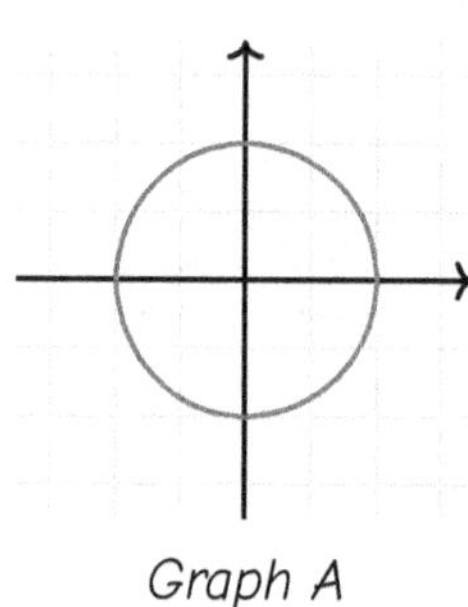

Graph A

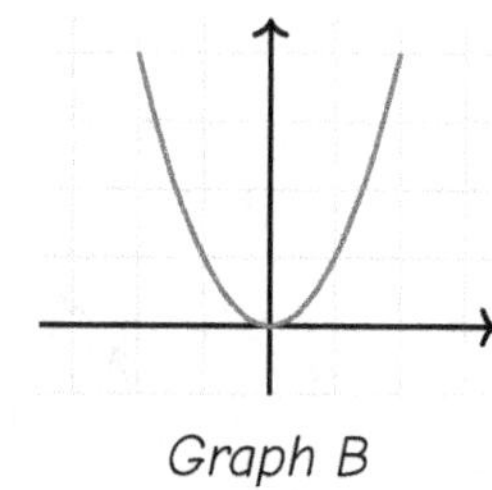

Graph B

(A) Graph A only

(B) Graph B only

(C) Both Graph A and Graph B

(D) Neither Graph A nor Graph B

12. If $f(x) = -x + 8$, what is $f(-3)$?

(A) 5

(B) -5

(C) 11

(D) -11

13. Company A charges \$20 plus \$8 per hour. Company B charges \$50 plus \$3 per hour. After how many hours do they charge the same amount?

Your Answer:

Find more at
ViewMath.com/SC-Grade8

ViewMath.com

14. Look at the graph below. Is the function linear or nonlinear? Explain your reasoning.

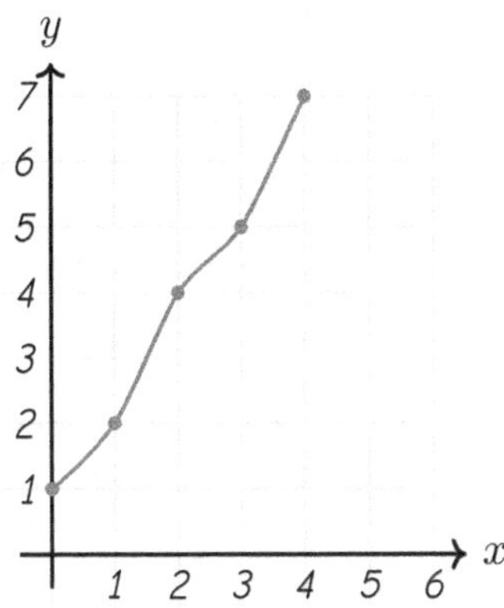

Your Answer

15. Find the equation of the line through $(1, 3)$ and $(4, 12)$.

Your Answer:

16. A graph of savings over time goes up steeply for months 1–3, then less steeply for months 4–6. What changed?

Your Answer

17. Point $N(0, 4)$ is rotated $180°$ around the origin. What are the coordinates of N'?

Your Answer

18. Triangle RST has vertices $R(0,0)$, $S(4,0)$, and $T(0,3)$. Triangle $R'S'T'$ has vertices $R'(0,0)$, $S'(0,4)$, and $T'(-3,0)$. What transformation maps RST to $R'S'T'$?

(A) Translation

(B) Reflection over the x-axis

(C) $90°$ counterclockwise rotation

(D) $90°$ clockwise rotation

19. Dilate the point $(-4, 10)$ by a factor of $\frac{1}{2}$ from the origin. What is the image?

Your Answer:

20. Square $ABCD$ is dilated from the origin by $k = 2$ to form $A'B'C'D'$. Which graph shows the correct image?

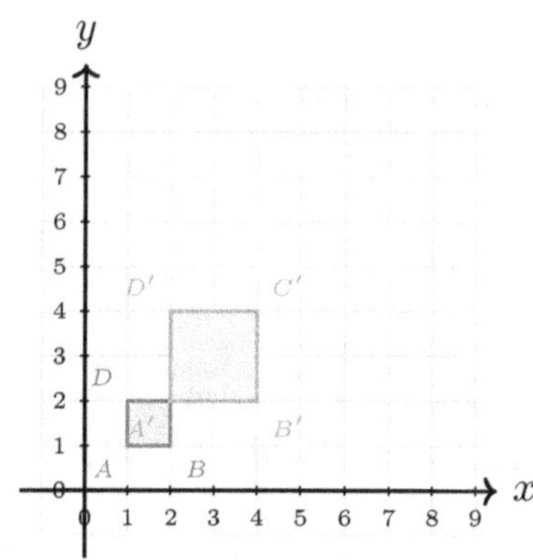

(A) $A'(2,2)$, $B'(4,2)$, $C'(4,4)$, $D'(2,4)$ — as shown

(B) $A'(3,3)$, $B'(4,3)$, $C'(4,4)$, $D'(3,4)$

(C) $A'(1,1)$, $B'(3,1)$, $C'(3,3)$, $D'(1,3)$

(D) $A'(2,1)$, $B'(4,1)$, $C'(4,2)$, $D'(2,2)$

21. A co-interior angle measures $97°$. What is the other co-interior angle?

Your Answer:

22. Which equation represents the Pythagorean Theorem?

(A) $a + b = c$ (B) $a^2 + b^2 = c^2$

(C) $a^2 \times b^2 = c^2$ (D) $2a + 2b = c$

23. What is the distance between $(1, 3)$ and $(4, 7)$?

(A) 7   (B) 25

(C) $\sqrt{7}$ (D) 5

24. Find the volume of a sphere with radius 6 in. Leave your answer in terms of π.

Your Answer:

25. Can a scatter plot show a positive association that is nonlinear? Give an example.

Your Answer:

26. True or false: The line of best fit must pass through at least one data point.

(A) True — it must go through the middle point.   (B) True — it must go through the first point.

(C) False — it does not need to go through any   (D) False — it must stay above all data points.
data point.

27. A model $y = -2x + 50$ represents the number of pages left to read (y) after x hours. How long until the book is finished (all pages read)?

Your Answer:

Find more at
ViewMath.com/SC-Grade8

28.

	Bike	Bus	Total
6th grade	18	22	40
7th grade	25	15	40
Total	43	37	80

What percentage of 6th graders ride the bus?

(A) 45%

(B) 55%

(C) 27.5%

(D) 62.5%

29. Data: $\{50, 55, 60, 65, 70\}$. Calculate the MAD.

Your Answer

30. $P(event) = \frac{3}{4}$. What is $P(not\ event)$?

(A) $\frac{3}{4}$

(B) $\frac{1}{4}$

(C) $\frac{1}{2}$

(D) 0

 *End of Practice Test 8* 

Great job finishing the test!

 *My Score*

I got ____________ out of 30 questions right.

*Check your answers in the **Answer Key** at the back of the book.*

Review any questions you missed. That's how we learn!

📊 Check Your Score Online!

Visit **ViewMath Academy** to enter your answers and see which topics
you need to review. You can also explore lessons, take quizzes, track
your scores, and save your progress!

viewmath.com/score/8.1.SC.23

Or go to viewmath.com/score and enter code: 8.1.SC.23

Practice Test 9

 30 Questions

✏️ Before You Start ✏️

- ✔ **Read each question carefully** before choosing your answer.
- ✔ **Show your work** on scratch paper when you need to.
- ✔ **Skip hard questions** and come back to them later.
- ✔ **Check your answers** when you're done.
- ✔ **Take your time** — there's no rush!

 ⭐ You've Got This! ⭐

Do your best and show what you know!

1. Which of the following could NOT be the decimal expansion of a rational number?

(A) 2.500

(B) $0.\overline{9}$

(C) $1.41421356\ldots$ (non-repeating)

(D) $0.\overline{36}$

2. What is $0.\overline{4}$ written as a fraction?

(A) $\frac{4}{10}$

(B) $\frac{4}{9}$

(C) $\frac{2}{5}$

(D) $\frac{1}{4}$

3. A student says $\sqrt{45}$ is between 6 and 7. Is the student correct?

(A) Yes, because $6^2 = 36$ and $7^2 = 49$.

(B) No, it is between 5 and 6.

(C) No, it is between 7 and 8.

(D) No, $\sqrt{45} = 9$ exactly.

4. Which is larger: $\sqrt{2} + \sqrt{3}$ or $\sqrt{10}$?

(A) $\sqrt{2} + \sqrt{3}$, because $2 + 3 = 5 > \sqrt{10}$

(B) $\sqrt{10}$, because $10 > 5$

(C) $\sqrt{10}$, because $\sqrt{10} \approx 3.16 > 3.15 \approx \sqrt{2} + \sqrt{3}$

(D) They are equal.

5. Between which two consecutive whole numbers does $\sqrt{50}$ lie?

(A) 6 and 7

(B) 7 and 8

(C) 24 and 26

(D) 8 and 9

6. Which is larger: 8×10^5 or 3×10^6?

(A) 8×10^5, because $8 > 3$

(B) 3×10^6, because the exponent is larger

(C) They are equal

(D) It cannot be determined

Find more at
ViewMath.com/SC-Grade8

7. Simplify $(7 \times 10^{-3})(8 \times 10^{-2})$. Write in proper scientific notation.

(A) 56×10^{-5}

(B) 5.6×10^{-5}

(C) 5.6×10^{-4}

(D) 5.6×10^{6}

8. Two delivery services are compared. Service P delivers $y = 4x$ packages per hour. Service Q's data is shown in the table.

Hours (x)	2	4	6	8
Packages (y)	10	20	30	40

How many more packages does Service Q deliver than Service P in 10 hours?

Your Answer

9. Find the slope through $(0, -2)$ and $(3, 7)$.

(A) 3

(B) -3

(C) $\frac{5}{3}$

(D) $\frac{9}{3}$

10. A store sells T-shirts for \$12 and hats for \$8. Use the information in the table to find how many of each were sold.

	Value
Total items sold	15
Total revenue	\$148

Your Answer

Find more at
ViewMath.com/SC-Grade8

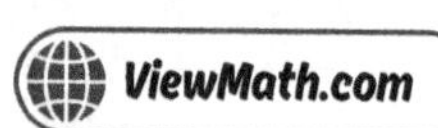

11. Which of these real-world examples is a function?

(A) A person's age mapped to their height (a person can be the same height at different ages)

(B) A student's name mapped to their student ID number

(C) A city mapped to all the people who live there

(D) A month mapped to every holiday in that month

12. If $f(x) = 3x + 1$, what is $f(4)$?

(A) 7

(B) 12

(C) 13

(D) 15

13. Two hikers start a trail. Hiker A walks 3 miles per hour and starts 2 miles from the trailhead. Hiker B is described by $y = 4x$. Who walks faster?

(A) Hiker A

(B) Hiker B

(C) They walk at the same speed.

(D) Cannot be determined.

14. Which is a characteristic of a linear function?

(A) Its graph is a curve.

(B) Its rate of change is constant.

(C) Its rate of change varies.

(D) It always passes through the origin.

15. A line passes through $(-1, 2)$ and $(3, 10)$. What is the slope?

(A) 2

(B) 3

(C) 4

(D) 8

16. A graph shows temperature over a day. From 6 AM to noon, the graph goes up steeply. From noon to 3 PM, it goes up slowly. What can you conclude?

(A) The temperature decreased between noon and 3 PM.

(B) The temperature increased faster in the morning than in the afternoon.

(C) The temperature was constant from noon to 3 PM.

(D) The temperature increased at the same rate all day.

17. Point $P(2,0)$ is rotated 90° counterclockwise around the origin. Where does P' land?

(A) $(0,-2)$

(B) $(0,2)$

(C) $(-2,0)$

(D) $(2,0)$

18. Which pair of figures is NOT necessarily congruent?

(A) A triangle and its reflection

(B) A rectangle and its translation

(C) Two rectangles with the same perimeter

(D) A square and its 90° rotation

19. Point $A(0,-5)$ is rotated 90° counterclockwise around the origin. Where does A' land?

(A) $(5,0)$

(B) $(0,5)$

(C) $(-5,0)$

(D) $(0,-5)$

20. A triangle has an area of 20 cm^2. It is dilated by $k=3$. What is the area of the image?

(A) 60 cm^2

(B) 180 cm^2

(C) 120 cm^2

(D) 6.67 cm^2

Find more at
ViewMath.com/SC-Grade8

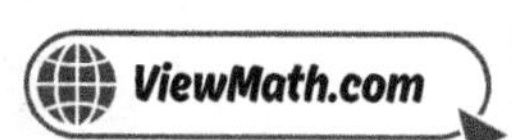

21. *Parallel lines ℓ and m are cut by transversal t. If a pair of alternate interior angles are $(4x)°$ and $80°$, what is x?*

(A) 320

(B) 20

(C) 25

(D) 10

22. *In a right triangle, the legs are 6 and 8. What is the hypotenuse?*

(A) 14

(B) 10

(C) 48

(D) $\sqrt{48}$

23. *What is the distance between $(0,0)$ and $(1,1)$?*

(A) 2

(B) 1

(C) $\sqrt{2}$

(D) 0

24. *What is the volume formula for a cone?*

(A) $V = \pi r^2 h$

(B) $V = \frac{1}{2}\pi r^2 h$

(C) $V = \frac{1}{3}\pi r^2 h$

(D) $V = \frac{4}{3}\pi r^3$

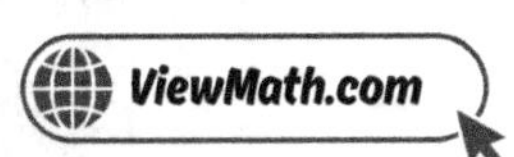

25. What type of association does this scatter plot show?

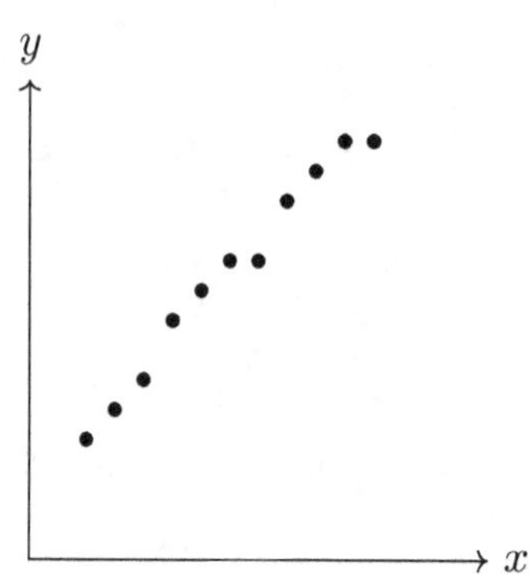

(A) Negative linear

(B) No association

(C) Positive linear

(D) Nonlinear

26. A trend line has equation $y = 1.5x + 2$. Using this, what is y when $x = 4$?

(A) 6

(B) 7

(C) 8

(D) 10

27. A scatter plot shows data and a trend line. The actual point at $x = 3$ is at $y = 8$, but the trend line passes through $(3, 6)$. What is the residual?

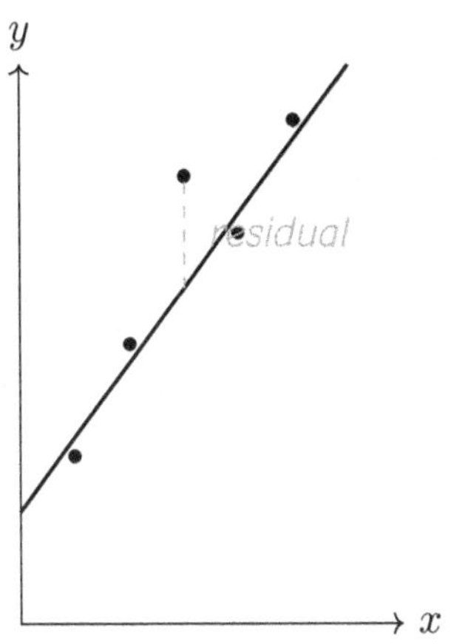

(A) -2

(B) 2

(C) 6

(D) 14

28. Using the same table, what percentage of students who did NOT study passed?

(A) 40%

(B) 60%

(C) 25%

(D) 50%

29. Data: $\{3, 5, 7, 9, 11, 13\}$. Mean $= 8$. What is the MAD?

(A) 2

(B) 3

(C) 4

(D) 8

30. Explain the difference between independent and dependent events with an example.

Your Answer:

 # End of Practice Test 9 

Great job finishing the test!

 My Score

I got _____________ out of 30 questions right.

*Check your answers in the **Answer Key** at the back of the book.*

Review any questions you missed. That's how we learn!

Check Your Score Online!

Visit **ViewMath Academy** to enter your answers and see which topics you need to review. You can also explore lessons, take quizzes, track your scores, and save your progress!

viewmath.com/score/8.1.SC.24

Or go to viewmath.com/score and enter code: 8.1.SC.24

Practice Test 10

30 Questions

✏️ Before You Start ✏️

- ✓ **Read each question carefully** before choosing your answer.
- ✓ **Show your work** on scratch paper when you need to.
- ✓ **Skip hard questions** and come back to them later.
- ✓ **Check your answers** when you're done.
- ✓ **Take your time** — there's no rush!

⭐ You've Got This! ⭐

Do your best and show what you know!

1. Look at the Venn diagram below. In which region does $\sqrt{5}$ belong?

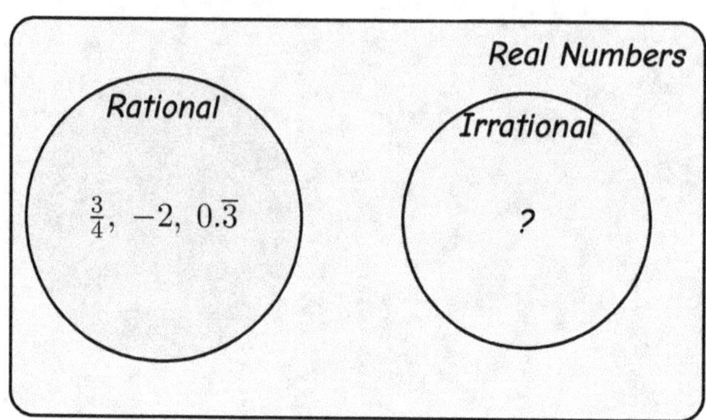

(A) Rational region, because $\sqrt{5} = 2.5$

(B) Irrational region, because 5 is not a perfect square

(C) Rational region, because $\sqrt{5}$ is a square root

(D) Neither region, because $\sqrt{5}$ is not a real number

2. The steps below show a student's work for converting $0.\overline{2}$ to a fraction. Which step contains the error?

Step	Work
1	Let $x = 0.222\ldots$
2	$10x = 2.222\ldots$
3	$10x - x = 2.222\ldots - 0.222\ldots$
4	$9x = 2$
5	$x = \frac{2}{10}$

(A) Step 2

(B) Step 3

(C) Step 4

(D) Step 5

3. Between which two consecutive integers does $\sqrt{50}$ lie?

(A) 6 and 7

(B) 7 and 8

(C) 8 and 9

(D) 24 and 26

Find more at
ViewMath.com/SC-Grade8

ViewMath.com

4. *Order from least to greatest:* π, $\sqrt{10}$, 3.2.

(A) π, 3.2, $\sqrt{10}$

(B) 3.2, π, $\sqrt{10}$

(C) $\sqrt{10}$, π, 3.2

(D) π, $\sqrt{10}$, 3.2

5. *Which statement about* $\sqrt{2}$ *is true?*

(A) $\sqrt{2}$ *is a rational number*

(B) $\sqrt{2}$ *is an integer*

(C) $\sqrt{2}$ *is an irrational number*

(D) $\sqrt{2} = 1.5$

6. *The diagram below shows three cells viewed under a microscope, with their actual sizes labeled. Write all three sizes in scientific notation, then list them from smallest to largest.*

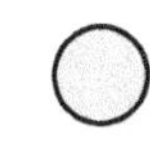

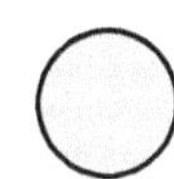

Cell A
0.008 mm

Cell B
0.05 mm

Cell C
0.0003 mm

Your Answer:

7. *Compute* $\frac{3.6 \times 10^{10}}{1.2 \times 10^4}$. *Write your answer in scientific notation.*

Your Answer:

8. *If* $y = kx$ *and* $y = 54$ *when* $x = 6$, *find* y *when* $x = 11$.

Your Answer:

9. A line goes down from left to right. Its slope must be:

(A) Positive

(B) Negative

(C) Zero

(D) Undefined

10. Pens cost \$2 and notebooks cost \$5. You buy 8 items for \$25. How many notebooks did you buy?

(A) 2

(B) 3

(C) 4

(D) 5

11. Use the mapping diagram below. List all the inputs that are part of this relation.

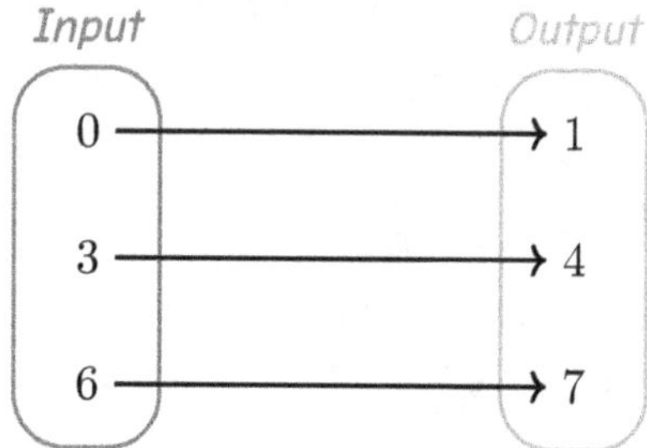

Your Answer:

12. If $f(x) = 6x - 4$, for which value of x is $f(x) = 0$?

(A) $x = 0$

(B) $x = \frac{2}{3}$

(C) $x = 1$

(D) $x = 4$

Find more at
ViewMath.com/SC-Grade8

13. The graph below shows Function F. Function G is $y = 3x + 2$. Find the rate of change of each function and state which is greater.

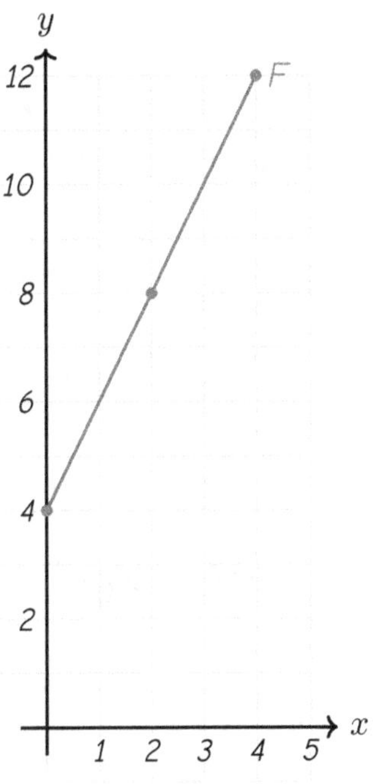

Your Answer

14. What is the form of a linear function?

(A) $y = ax^2 + bx + c$ (B) $y = mx + b$

(C) $y = a^x$ (D) $y = \frac{a}{x}$

15. Sam has \$120 and spends \$15 per day. Which function models the money remaining after x days?

(A) $y = 15x + 120$ (B) $y = -15x + 120$

(C) $y = 120x - 15$ (D) $y = -120x + 15$

16. A graph of amount of gas in a car over time goes downward. Is the function increasing, decreasing, or constant?

Your Answer:

17. A student says that a reflection changes the area of a figure. Is this correct?

(A) Yes, because the figure flips.

(B) Yes, because one dimension reverses.

(C) No, because reflections preserve all measure-ments.

(D) No, but reflections change the perimeter.

18. Triangle ABC is shown on the grid below. $A'B'C'$ is congruent to ABC and formed by reflecting ABC over the x-axis. What are the coordinates of C'?

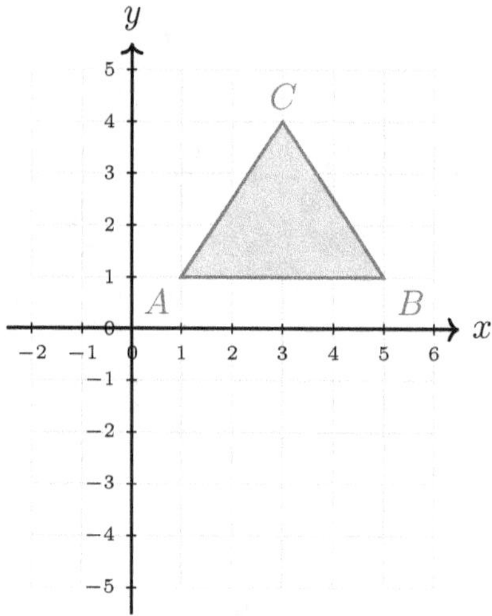

Your Answer:

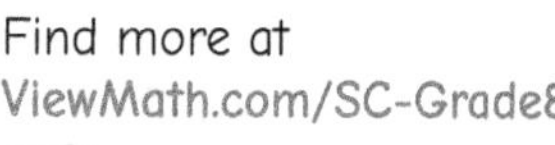
Find more at
ViewMath.com/SC-Grade8

19. A point is translated by $(3, -2)$ and then reflected over the y-axis. If the original point is $(1, 4)$, what is the final image?

(A) $(-4, 2)$

(B) $(4, 2)$

(C) $(-4, -2)$

(D) $(4, -2)$

20. A figure is dilated by $k = 0.25$. What happens to the figure?

(A) It gets 4 times larger.

(B) It stays the same size.

(C) It shrinks to $\frac{1}{4}$ of its original size.

(D) It disappears.

21. What is the sum of the interior angles of any triangle?

(A) $90°$

(B) $180°$

(C) $270°$

(D) $360°$

22. A right triangle has hypotenuse 17 and one leg 8. What is the other leg?

(A) 9

(B) 15

(C) 25

(D) $\sqrt{225}$

23. A rectangle has corners at $(0, 0)$ and $(9, 12)$. What is the length of the diagonal?

Your Answer:

Find more at
ViewMath.com/SC-Grade8

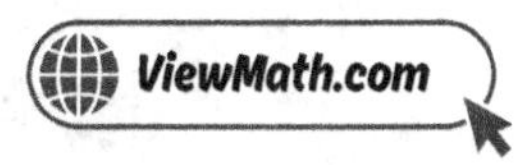

24. A sphere has the dimensions shown. What is its volume? Use $\pi \approx 3.14$.

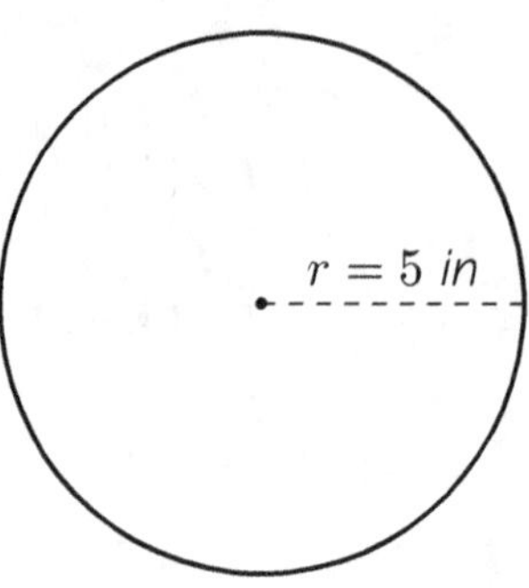

(A) $78.5\ in^3$

(B) $523.3\ in^3$

(C) $392.5\ in^3$

(D) $314.0\ in^3$

25. True or false: A scatter plot can show both clustering and outliers at the same time.

(A) True — they describe different features of the data.

(B) False — data has either clusters or outliers, not both.

(C) True — but only if there is no association.

(D) False — outliers are always in clusters.

26. Why should outliers generally be ignored when drawing a line of best fit?

Your Answer

27. A linear model is $y = 3x + 10$, where x is hours studied and y is the test score. What does the y-intercept represent?

(A) The score after 3 hours of studying

(B) The score with 0 hours of studying

(C) The number of hours needed to pass

(D) The maximum possible score

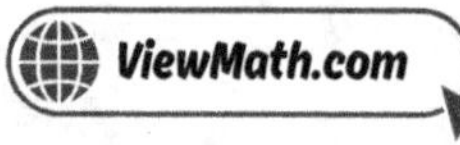

28. The bar chart below shows survey results about favorite season by gender. Which statement is supported?

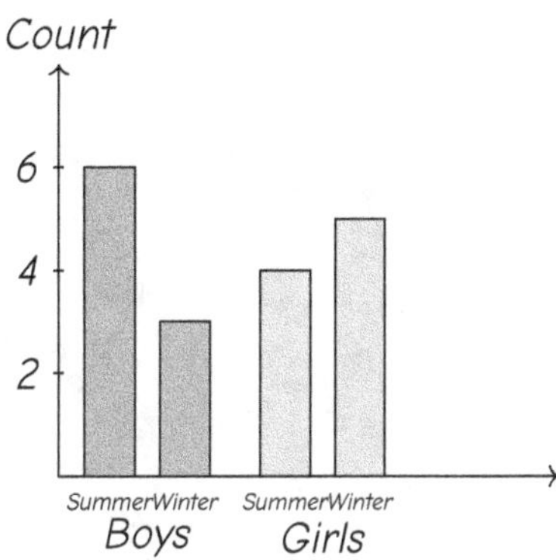

(A) Boys and girls prefer summer equally.

(B) Boys prefer summer more, girls prefer winter more.

(C) Girls prefer summer more than boys.

(D) There is no difference in preference.

29. If every value in a data set is multiplied by 3, the MAD:

(A) stays the same

(B) is multiplied by 3

(C) is divided by 3

(D) is multiplied by 9

30. A bag has 5 red and 3 blue marbles. One marble is drawn and replaced. Then another is drawn. What is $P(\text{red, then blue})$?

(A) $\frac{15}{64}$

(B) $\frac{15}{56}$

(C) $\frac{5}{8}$

(D) $\frac{3}{8}$

 End of Practice Test 10 

Great job finishing the test!

 My Score

I got __________ out of 30 questions right.

*Check your answers in the **Answer Key** at the back of the book.*

💡 *Review any questions you missed. That's how we learn!*

📊 Check Your Score Online!

Visit **ViewMath Academy** to enter your answers and see which topics you need to review. You can also explore lessons, take quizzes, track your scores, and save your progress!

viewmath.com/score/8.1.SC.25

Or go to viewmath.com/score and enter code: 8.1.SC.25

Answer Key & Explanations

Answer Key

First try each test on your own, then check your work here.

☑ Practice Test 1 — Answer Key

1 False **2** C **3** D **4** B **5** B **6** $\approx 3.1 \times 10^{12}$ **7** C **8** B **9** 3

10 80 **11** C **12** 3 **13** Yes **14** C **15** 4 **16** A

17 Translation, reflection, rotation **18** 55° **19** B **20** B **21** C **22** B **23** A

24 C **25** B **26** A **27** B

28 Boys: $\frac{20}{30} \approx 66.7\%$. Girls: $\frac{12}{20} = 60\%$. The percentages are close, so association is weak or none. **29** B

30 $\frac{4}{52} \times \frac{3}{51} = \frac{12}{2652} = \frac{1}{221}$

💡 Time to Learn! 💡

*Review the explanations below, **especially for the questions you missed**.*

Understanding why each answer is correct builds stronger problem-solving skills.

Tip: *Circle any questions you got wrong, then read their explanation carefully.*

📖 Practice Test 1 — Detailed Explanations

Find more at
ViewMath.com/SC-Grade8

1. $\frac{22}{7} \approx 3.142857\ldots$ is a rational approximation of π, but $\pi = 3.14159265\ldots$ is irrational. They are close but not equal.

2. $0.\overline{123}$ has a 3-digit repeating block, so you should multiply by $10^3 = 1000$, not 100.

3. $3.74^2 = 13.9876$ (under) and $3.75^2 = 14.0625$ (just over). 14.0625 is very close to 14, so 3.75 is the best next estimate.

4. $\sqrt{7} \approx 2.646$, so $5 - \sqrt{7} \approx 5 - 2.646 = 2.354 \approx 2.4$.

5. If $s = 8$, then $A = 8^2 = 64$. If $e = 4$, then $V = 4^3 = 64$. Since $A = V = 64$, the pair $s = 8, e = 4$ works.

6. $498{,}000{,}000 \approx 5 \times 10^8$ and $6{,}200 \approx 6.2 \times 10^3$. Product $\approx 5 \times 6.2 \times 10^{11} = 31 \times 10^{11} = 3.1 \times 10^{12}$.

7. $(3 \times 10^4)(5 \times 10^3) = 15 \times 10^7 = 1.5 \times 10^8$.

8. At \$8 per hour with no extra fee, the relationship is proportional: $y = 8x$.

9. $m = \frac{11-(-1)}{7-3} = \frac{12}{4} = 3$.

10. $a + s = 200$ and $8a + 5s = 1240$. From first: $s = 200 - a$. Substitute: $8a + 5(200 - a) = 1240$, so $3a + 1000 = 1240$, $3a = 240$, $a = 80$.

11. At $x = 3$ there are two different y-values (1 and 5). A vertical line at $x = 3$ would cross two points, so this is not a function.

12. Follow the dashed line at $y = 6$ until it meets the graph. That point is $(3, 6)$, so $x = 3$.

13. Lines with the same slope but different y-intercepts are parallel.

Find more at
ViewMath.com/SC-Grade8

14 A linear function has the form $y = mx + b$. Only $y = 4x - 7$ fits this form with $m = 4$ and $b = -7$.

15 $\frac{25-9}{6-2} = \frac{16}{4} = 4$.

16 Walking increases distance. Waiting means distance stays the same (constant). Walking again increases distance further.

17 The three rigid transformations are translation (slide), reflection (flip), and rotation (turn). They all preserve size and shape.

18 $\angle R = 180 - 72 - 53 = 55°$. Since $\angle R$ corresponds to $\angle U$, $\angle U = 55°$.

19 Original sides: 6, 8, and $\sqrt{6^2 + 8^2} = 10$. Perimeter $= 6 + 8 + 10 = 24$. Dilation by $\frac{1}{2}$ multiplies every length by $\frac{1}{2}$, so the new perimeter is $24 \times \frac{1}{2} = 12$.

20 Divide the image coordinates by the original: $\frac{12}{4} = 3$ and $\frac{-6}{-2} = 3$. The scale factor is 3.

21 Corresponding angles (same position at each intersection) are equal when the lines are parallel.

22 $c^2 = 5^2 + 5^2 = 25 + 25 = 50$, so $c = \sqrt{50}$.

23 $d = \sqrt{(4-(-2))^2 + (9-1)^2} = \sqrt{6^2 + 8^2} = \sqrt{36 + 64} = \sqrt{100} = 10$.

24 The sphere formula $V = \frac{4}{3}\pi r^3$ uses only the radius. There is no separate height for a sphere.

25 One variable goes up while the other goes down, forming a negative association.

26 Slope $= \frac{7-3}{5-1} = 1$. y-intercept: $3 = 1(1) + b$, $b = 2$. Equation: $y = x + 2$.

27 The linear pattern observed in data may not hold outside the data range. Real-world relationships can change.

28 Compare the conditional relative frequencies. Since 66.7% and 60% are similar, the preference doesn't change much by gender.

29 $MAD = 0$ means every deviation is 0, so every data value equals the mean.

30 First ace: $\frac{4}{52}$. Second ace (without replacement): $\frac{3}{51}$. Multiply: $\frac{12}{2652} = \frac{1}{221}$.

✅ Practice Test 2 — Answer Key

1 C

2 Multiply by 100; $100x = 54.\overline{54}$; $99x = 54$; $x = \frac{54}{99} = \frac{6}{11}$

3 A: ≈ 4.2, B: ≈ 6.3, C: ≈ 8.5

4 C **5** B **6** B **7** A **8** 180 mL **9** B **10** 5 **11** C **12** 3

13 Function A: rate = 3, initial = 20. Function B: rate = 5, initial = 5. At $x = 10$, $B = 55$ is greater than $A - 50$.

14 $y = x^2$ **15** B **16** C **17** C **18** B **19** $(1, 2)$ **20** C **21** A **22** No

23 B **24** $100\pi\ cm^3$ **25** C **26** Approximate line through $(1, 2)$ and $(7, 9)$: slope $\approx \frac{7}{6} \approx 1.17$.

27 B **28** D **29** A **30** C

💡 Time to Learn! 💡

Review the explanations below, **especially for the questions you missed**.

Understanding why each answer is correct builds stronger problem-solving skills.

Tip: Circle any questions you got wrong, then read their explanation carefully.

Find more at
ViewMath.com/SC-Grade8

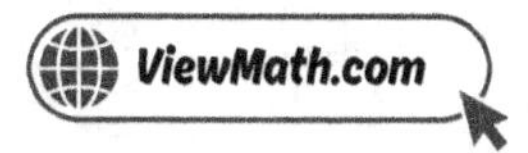

📖 Practice Test 2 — Detailed Explanations

1. $P = \sqrt{2}$ is irrational (since 2 is not a perfect square) and $Q = 3$ is an integer, which is rational.

2. The repeating block has 2 digits, so multiply by 100. Then $100x - x = 99x = 54$, giving $x = \frac{54}{99} = \frac{6}{11}$.

3. Side $= \sqrt{Area}$. $\sqrt{18} \approx 4.24 \approx 4.2$. $\sqrt{40} \approx 6.32 \approx 6.3$. $\sqrt{72} \approx 8.49 \approx 8.5$.

4. $\sqrt{5} \approx 2.236$ and $\sqrt{3} \approx 1.732$. Adding: $2.236 + 1.732 = 3.968 \approx 4.0$.

5. $x = \pm\sqrt{\frac{9}{16}} = \pm\frac{3}{4}$, since $\left(\frac{3}{4}\right)^2 = \frac{9}{16}$.

6. $\frac{4.2 \times 10^3}{4.2 \times 10^{-3}} = 10^{3-(-3)} = 10^6$. So 4.2×10^3 is 10^6 (one million) times as large.

7. Divide the coefficients: $\frac{8}{2} = 4$. Subtract the exponents: $10^{7-3} = 10^4$. Answer: 4×10^4.

8. Rate $= \frac{45}{15} = 3$ mL/min. In 60 minutes: $3 \times 60 = 180$ mL.

9. Rate of change $= \frac{16-4}{6} = \frac{12}{6} = 2$ cm per week.

10. $a + b = 12$ and $5a + 10b = 85$. From first: $a = 12 - b$. Substitute: $5(12 - b) + 10b = 85$, so $60 + 5b = 85$, $5b = 25$, $b = 5$.

11. Find the ordered pair with input 0: $(0, 1)$. The output is 1.

12. $f(2) - f(1) = 7 - 4 = 3$.

Find more at
ViewMath.com/SC-Grade8

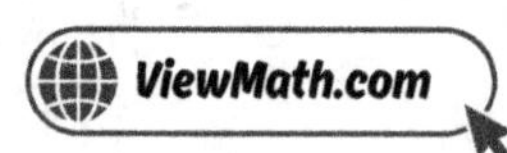

13 A: slope 3, $b = 20 \Rightarrow A(10) = 3(10) + 20 = 50$. B: slope 5, $b = 5 \Rightarrow B(10) = 5(10) + 5 = 55$. Function B is greater at $x = 10$.

14 Any equation where x has an exponent other than 1 (such as x^2, x^3, $\frac{1}{x}$, $\sqrt{x}$) is nonlinear.

15 Slope $= \frac{18-6}{4-1} = \frac{12}{3} = 4$.

16 Filling: water level rises (increasing). Soaking: level stays the same (constant). Draining: level drops (decreasing).

17 Reflecting over the y-axis flips the sign of the x-coordinate while keeping y the same: $(7, -2) \rightarrow (-7, -2)$.

18 When triangles are congruent, all corresponding sides are equal. Since AB corresponds to DE, $DE = 8$ cm.

19 Add the translation: $(4 + (-3), -6 + 8) = (1, 2)$.

20 Check ratios: $\frac{6}{3} = 2$ but $\frac{9}{5} = 1.8$. Since $2 \neq 1.8$, the sides are not proportional and the rectangles are not similar.

21 The exterior angle adjacent to the third angle equals the sum of the two non-adjacent interior angles: $37 + 53 = 90°$.

22 $5^2 + 10^2 = 25 + 100 = 125$, but $12^2 = 144$. Since $125 \neq 144$, it is not a right triangle.

23 The points are on the x-axis, so the distance is just the horizontal difference: $|5 - 0| = 5$.

24 $V = \frac{1}{3}\pi(25)(12) = \frac{300\pi}{3} = 100\pi$ cm^3.

25 Point P at $(3, 7)$ is far above the general upward trend around $y \approx 2.5$ when $x = 3$. It is the outlier.

Find more at
ViewMath.com/SC-Grade8

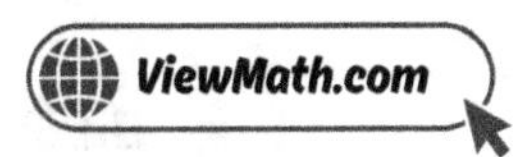

26. $m = \frac{9-2}{7-1} = \frac{7}{6} \approx 1.17$. The line rises about 1.2 units for each 1-unit increase in x.

27. In $y = mx + b$, the slope m represents the rate of change: how much y changes for each 1-unit increase in x.

28. Joint frequencies fill the interior cells: $3 \times 4 = 12$.

29. Deviations: $|3-6| = 3, |5-6| = 1, |5-6| = 1, |7-6| = 1, |10-6| = 4$. Sum $= 10$. MAD should be checked: mean $= (3+5+5+7+10)/5 = 6$. MAD $= 10/5 = 2$. Actually 2.

30. 2 coin outcomes $\times$ 6 die outcomes $= 12$ total outcomes.

✅ Practice Test 3 — Answer Key

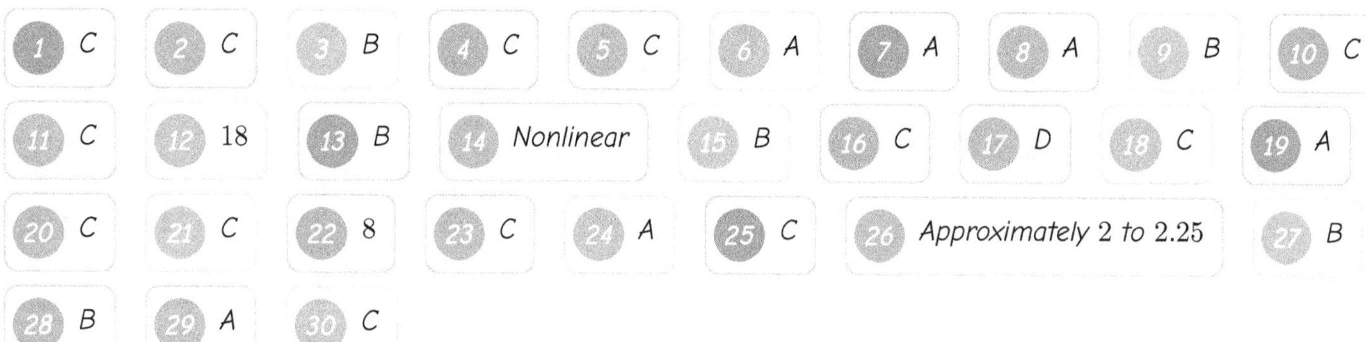

1	C	2	C	3	B	4	C	5	C	6	A	7	A	8	A	9	B	10	C
11	C	12	18	13	B	14	Nonlinear	15	B	16	C	17	D	18	C	19	A
20	C	21	C	22	8	23	C	24	A	25	C	26	Approximately 2 to 2.25	27	B
28	B	29	A	30	C

💡 Time to Learn! 💡

Review the explanations below, **especially for the questions you missed**.

Understanding why each answer is correct builds stronger problem-solving skills.

Tip: Circle any questions you got wrong, then read their explanation carefully.

📖 Practice Test 3 — Detailed Explanations

1 $0.\overline{81}$ is a repeating decimal, so it can be written as a fraction: $0.\overline{81} = \frac{81}{99} = \frac{9}{11}$. The others are irrational.

2 Let $x = 0.\overline{27}$. Then $100x = 27.\overline{27}$. Subtract: $99x = 27$, so $x = \frac{27}{99} = \frac{3}{11}$.

3 $7.7^2 = 59.29$ and $7.8^2 = 60.84$. So $\sqrt{60} \approx 7.75$, making 8 a less accurate estimate. 7.7 or 7.8 is better.

4 $4\sqrt{2} \approx 4 \times 1.414 = 5.656$ and $\sqrt{30} \approx 5.477$. Since $5.656 > 5.477$, $4\sqrt{2}$ is larger.

5 Side $= \sqrt{196} = 14$ cm, since $14 \times 14 = 196$.

6 Move the decimal 4 places right: $0.00072 = 7.2 \times 10^{-4}$.

7 $(9.5 \times 10^{-18})(2 \times 10^6) = 19 \times 10^{-12} = 1.9 \times 10^{-11}$ g.

8 Store A: $k = 2$ per pound. Store B: $k = \frac{7.50}{3} = 2.50$ per pound. Store A is cheaper.

9 $\frac{270-120}{5-2} = \frac{150}{3} = 50$ mph.

10 $60t + 80t = 420$, so $140t = 420$, $t = 3$ hours.

11 In a function, the first coordinate x is the input and the second coordinate y is the output.

12 $f(3) = 4(3) + 1 = 13$. $f(1) = 4(1) + 1 = 5$. Sum: $13 + 5 = 18$.

Find more at
ViewMath.com/SC-Grade8

13 Function A earns \$12/week ($m = 12$). Function B earns \$15/week ($m = 15$). Since $15 > 12$, Function B earns more per week.

14 The variable x is raised to the 3rd power. For linear, x must have exponent 1.

15 No activation fee means $b = 0$. The monthly cost is \$35, so $m = 35$. The function is $y = 35x$.

16 The four sections are: increasing, decreasing, increasing, constant.

17 $(3, 2) \to (-3, -2)$ follows the rule $(x, y) \to (-x, -y)$, which is a $180°$ rotation around the origin.

18 A dilation by a factor other than 1 changes the size of the figure, so it does not preserve congruence.

19 Apply the rule: $(-1, 6) \to (6, -(-1)) = (6, 1)$.

20 All squares have four $90°$ angles. Any two squares can be mapped by a dilation since the ratio of their sides is constant. So all squares are similar.

21 The exterior angle equals the sum of the two non-adjacent interior angles, so it's greater than either one of them individually.

22 $x^2 = 17^2 - 15^2 = 289 - 225 = 64$, so $x = 8$.

23 The side length is 6. The diagonal of a square with side s is $s\sqrt{2} = 6\sqrt{2} \approx 8.49$.

24 Radius $= 12$. $V = \frac{4}{3}(3.14)(12^3) = \frac{4}{3}(3.14)(1728) = \frac{4}{3}(5425.92) \approx 7234.6$ cm^3.

25 A strong positive trend corresponds to an r-value close to 1. $r = 0.9$ indicates a strong positive association.

Find more at
ViewMath.com/SC-Grade8

26. Using endpoints: $m \approx \frac{11-2}{5-1} = \frac{9}{4} = 2.25$. A slope around 2 is a reasonable estimate.

27. Residual = actual − predicted. A positive residual means the actual value is above the predicted value.

28. The 20 is the count for a specific combination of two variables (7th grade AND soccer), making it a joint frequency.

29. $|2-6| = 4, |4-6| = 2, |6-6| = 0, |8-6| = 2, |10-6| = 4$

30. $P(6) = \frac{1}{6}$. Independent: $P = \frac{1}{6} \times \frac{1}{6} = \frac{1}{36}$.

☑ Practice Test 4 — Answer Key

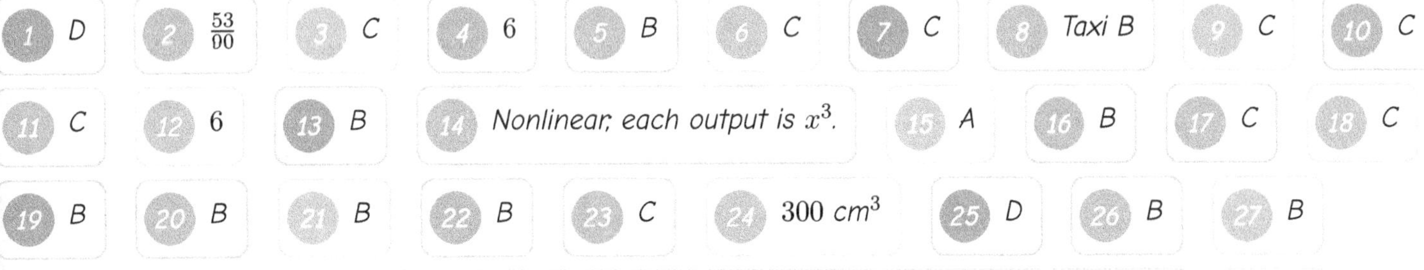

1	D	2	$\frac{53}{00}$	3	C	4	6	5	B	6	C	7	C	8	Taxi B	9	C	10	C
11	C	12	6	13	B	14	Nonlinear; each output is x^3.	15	A	16	B	17	C	18	C
19	B	20	B	21	B	22	B	23	C	24	300 cm^3	25	D	26	B	27	B
28	B	29	Set X: MAD $= 0$ (all values equal). Set Y: MAD $= 2.4$. Set Y has more variability.	30	A

💡 Time to Learn! 💡

Review the explanations below, **especially for the questions you missed**.

Understanding why each answer is correct builds stronger problem-solving skills.

Tip: Circle any questions you got wrong, then read their explanation carefully.

Find more at
ViewMath.com/SC-Grade8

📖 *Practice Test 4 — Detailed Explanations*

1. $\sqrt{7}$ is irrational because 7 is not a perfect square. The other choices are all rational: $\frac{5}{8} = 0.625$, $0.\overline{6} = \frac{2}{3}$, and $\sqrt{9} = 3$.

2. Let $x = 0.5888\ldots$. Then $10x = 5.888\ldots$ and $100x = 58.888\ldots$. Subtract: $90x = 53$, so $x = \frac{53}{90}$.

3. $3^2 = 9$ and $4^2 = 16$. Since $9 < 15 < 16$, we have $3 < \sqrt{15} < 4$. Therefore $\sqrt{15} > 3$ is true. In fact, $\sqrt{15} \approx 3.87$.

4. $\sqrt{2} \times \sqrt{18} = \sqrt{2 \times 18} = \sqrt{36} = 6$.

5. $7 \times 7 = 49$, so $\sqrt{49} = 7$.

6. Move the decimal 3 places right: $6.1 \times 10^3 = 6{,}100$.

7. $\frac{2 \times 10^{-6}}{10^{-6}} = 2$. The bacterium is 2 micrometers long.

8. Taxi A: \$3/mile. Taxi B: $\frac{36}{9} = \$4$/mile. Taxi B is more expensive per mile.

9. $m = \frac{5 - (-1)}{2 - (-4)} = \frac{6}{6} = 1$.

10. $x + y = 25$ and $x - y = 7$. Add: $2x = 32$, $x = 16$. The larger number is 16.

11. For any input x, there is exactly one output $y = -4x + 9$. Having a negative slope or negative outputs does not disqualify a function.

12. $8x - 6 = 42 \Rightarrow 8x = 48 \Rightarrow x = 6$.

Find more at
ViewMath.com/SC-Grade8

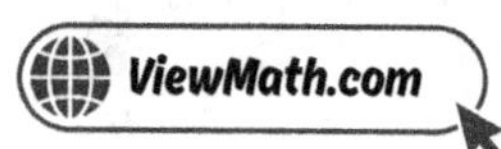

13. $P(0) = 5$. For Function Q, $b = 9$. Since $9 > 5$, Function Q has the greater initial value.

14. $1^3 = 1$, $2^3 = 8$, $3^3 = 27$, $4^3 = 64$. The differences $(7, 19, 37)$ are not constant, confirming nonlinear.

15. The slope $m = 50$ represents the amount added per week, so the weekly deposit is $50.

16. Graph B shows a steep increase (walking quickly), a flat section (waiting), then a gentler decrease (walking slowly home). Graph A's increase and decrease have similar steepness, so it doesn't match "quickly" then "slowly."

17. A rotation turns (spins) a figure around a fixed point called the center of rotation.

18. Two circles with the same radius are always congruent — one can be translated to overlap the other.

19. $180°$ rotation: $(a, b) \rightarrow (-a, -b)$. Reflect over x-axis: $(-a, -b) \rightarrow (-a, b)$.

20. When $k = 1$, every point stays in the same place. The image is congruent (identical in size and shape) to the original.

21. Alternate interior angles formed by parallel lines and a transversal are equal: $72°$.

22. $h^2 = 15^2 - 9^2 = 225 - 81 = 144$, so $h = 12$ ft.

23. The right triangle has legs 8 and 6. Distance $= \sqrt{8^2 + 6^2} = \sqrt{64 + 36} = \sqrt{100} = 10$.

24. A cylinder is 3 times the volume of a cone with the same base and height: $100 \times 3 = 300$ cm^3.

25. If there is no clear trend or pattern, there is no association between the variables.

26 $m = \frac{19-7}{9-3} = \frac{12}{6} = 2.$

27 Predicted: $y = 2.5(4) + 1 = 11.$ Residual $=$ actual $-$ predicted $= 12 - 11 = 1.$

28 Total who passed $= 40.$ Total students $= 60.$ Fraction $= \frac{40}{60} = \frac{2}{3}.$

29 Both have mean $= 5.$ Set X deviations all $0.$ Set Y deviations: $4, 2, 0, 2, 4,$ MAD $= 2.4.$

30 With replacement: $\frac{6}{10} \times \frac{6}{10} = \frac{36}{100} = \frac{9}{25}.$

✅ Practice Test 5 — Answer Key

1 Rational **2** B **3** A **4** B **5** C **6** 11 places to the left **7** A **8** D

9 B **10** Length $= 16$ m, width $= 8$ m **11** Yes **12** D **13** 7 **14** B **15** A

16 The speed is constant for 4 seconds. **17** A **18** C **19** C **20** A **21** 25 **22** B

23 A **24** B **25** C **26** B

27 At $x = 5$: $y = 30.$ At $x = 30$: $y = -45.$ The prediction at $x = 5$ is more reliable because it is within the data range.

28 B **29** Mean $= 5.$ Deviations: $4, 2, 0, 2, 4.$ MAD $= \frac{12}{5} = 2.4$ **30** B

Find more at
ViewMath.com/SC-Grade8

💡 **Time to Learn!** 💡

*Review the explanations below, **especially for the questions you missed**.*

Understanding why each answer is correct builds stronger problem-solving skills.

Tip: *Circle any questions you got wrong, then read their explanation carefully.*

📖 Practice Test 5 — Detailed Explanations

1. $\sqrt{144} = 12$ exactly, because $12^2 = 144$. Since 12 is an integer, $\sqrt{144}$ is rational.

2. The repeating block has 2 digits, so multiply by $10^2 = 100$. Then $100x = 63.6363\ldots$ and $x = 0.6363\ldots$. Subtracting: $100x - x = 63$.

3. $4^2 = 16$ and $5^2 = 25$. Since $16 < 22 < 25$, we have $4 < \sqrt{22} < 5$.

4. Area $= \sqrt{20} \times \sqrt{8} = \sqrt{160}$. $\sqrt{160} \approx 12.65$ cm^2. (Note: $\sqrt{160} = 4\sqrt{10} \approx 12.65$.) Choices A and C confuse addition with multiplication.

5. $81 = 9^2$, so 81 is a perfect square.

6. The exponent is -11, so you move the decimal 11 places to the left from 2.5.

7. Same exponent, so add the coefficients: $5.2 + 3.8 = 9.0$. Answer: 9×10^6.

8. If the relationship passes through $(1, 5)$ and $(3, 15)$, then $k = 5$. But $5 \times 2 = 10 \neq 13$, so $(2, 13)$ does not fit.

9. $m = \frac{7-3}{\frac{3}{2}-\frac{1}{2}} = \frac{4}{1} = 4$.

Find more at
ViewMath.com/SC-Grade8

10 $2l + 2w = 48$ and $l = 2w$. Substitute: $2(2w) + 2w = 48$, $6w = 48$, $w = 8$. Then $l = 16$.

11 Every input $(3, 4, 5, 6)$ maps to exactly one output (8). Repeated outputs are allowed, so this is a function.

12 $f(0) = -2(0) + 10 = 0 + 10 = 10$.

13 The initial value is the output when $x = 0$, which is 7.

14 For a function to be linear, x must have an exponent of 1. Since x is squared in $y = 3x^2$, the function is nonlinear regardless of the coefficient.

15 Slope: $\frac{22-10}{5-2} = \frac{12}{3} = 4$. Using $(2, 10)$: $10 = 4(2) + b \Rightarrow b = 2$. So $y = 4x + 2$.

16 A flat section means the output (speed) stays the same — the object travels at a steady speed for those 4 seconds.

17 Add the translation to each coordinate: $(1 + 3, 2 + 1) = (4, 3)$.

18 Both reflection and translation are rigid transformations. Any sequence of rigid transformations preserves congruence.

19 The rule $(x, y) \rightarrow (-y, x)$ is a 90° counterclockwise rotation around the origin.

20 Congruent figures are a special case of similar figures with $k = 1$. All congruent figures are similar, but not all similar figures are congruent.

21 $(2x + 5) + 3x + (x + 25) = 180$. Combine: $6x + 30 = 180$, $6x = 150$, $x = 25$.

22 $7^2 + 24^2 = 49 + 576 = 625 = 25^2$. Since $a^2 + b^2 = c^2$, it is a right triangle.

Find more at
ViewMath.com/SC-Grade8

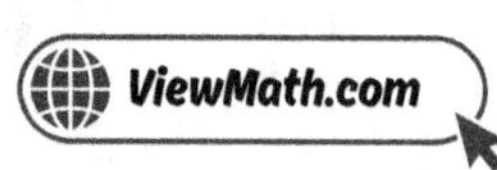

23. $d = \sqrt{(10-2)^2 + (9-3)^2} = \sqrt{64 + 36} = \sqrt{100} = 10.$

24. $V = \pi r^2 h = 3.14 \times 16 \times 10 = 502.4\ cm^3.$

25. A linear association means the data points roughly follow a straight-line trend.

26. A good fit means the data points are close to the line, with small distances from each point to the line.

27. $y = -3(5) + 45 = 30$ (interpolation). $y = -3(30) + 45 = -45$ (extrapolation — unreliable and gives a negative value that may not make sense).

28. $0.15 \times 200 = 30.$

29. Mean $= \frac{1+3+5+7+9}{5} = 5.$ Sum of absolute deviations $= 4 + 2 + 0 + 2 + 4 = 12.$ MAD $= 12 \div 5 = 2.4$

30. $P(A \text{ and } B) = 0.6 \times 0.4 = 0.24.$

✓ Practice Test 6 — Answer Key

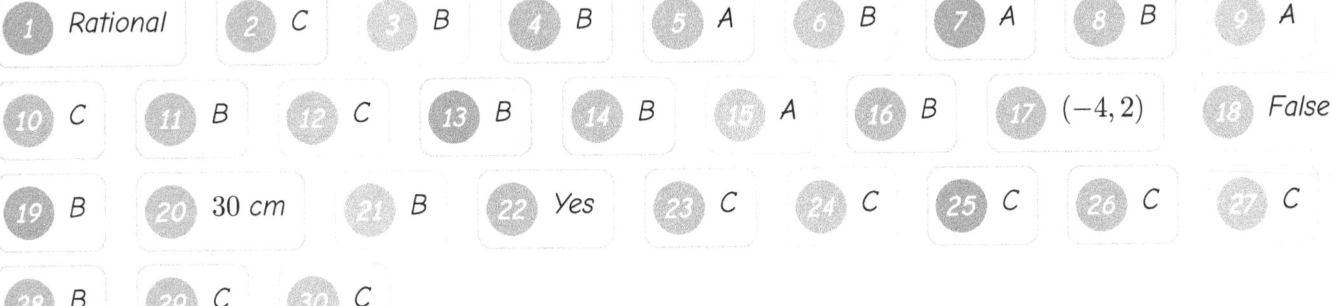

1 Rational	2 C	3 B	4 B	5 A	6 B	7 A	8 B	9 A
10 C	11 B	12 C	13 B	14 B	15 A	16 B	17 $(-4, 2)$	18 False
19 B	20 30 cm	21 B	22 Yes	23 C	24 C	25 C	26 C	27 C
28 B	29 C	30 C						

Find more at
ViewMath.com/SC-Grade8

> ## 💡 Time to Learn! 💡
>
> Review the explanations below, **especially for the questions you missed.**
>
> Understanding why each answer is correct builds stronger problem-solving skills.
>
> **Tip:** Circle any questions you got wrong, then read their explanation carefully.

📖 Practice Test 6 — Detailed Explanations

1. $\frac{5}{6}$ is a fraction of two integers with a nonzero denominator, so it is rational. Its decimal is $0.8\overline{3}$, which repeats.

2. The repeating block 123 has 3 digits, so you multiply by $10^3 = 1000$.

3. $10^2 = 100$ and $11^2 = 121$. Since $100 < 110 < 121$, we have $10 < \sqrt{110} < 11$.

4. $\sqrt{a} + \sqrt{b} \neq \sqrt{a+b}$. You must compute each root separately: $\sqrt{4} = 2$ and $\sqrt{9} = 3$, so the sum is 5. $\sqrt{13} \approx 3.6 \neq 5$.

5. Edge $= \sqrt[3]{343} = 7$ in., since $7^3 = 343$.

6. Move the decimal 5 places left: $9.03 \times 10^{-5} = 0.0000903$.

7. $4 \times (9.5 \times 10^{12}) = 38 \times 10^{12} = 3.8 \times 10^{13}$ km.

8. $k = \frac{y}{x} = \frac{9}{2} = 4.5$.

9. $m = \frac{1-7}{5-2} = \frac{-6}{3} = -2$.

Find more at
ViewMath.com/SC-Grade8

10. $n + d = 15$ and $5n + 10d = 110$. From first: $n = 15 - d$. Substitute: $5(15 - d) + 10d = 110$, so $75 + 5d = 110$, $5d = 35$, $d = 7$.

11. $y = 7$ is a horizontal line. Every x-value gives the same output 7, so it is a function. $x = 5$ is a vertical line (not a function). The other choices fail the vertical line test.

12. On the graph, go to $x = 2$ and read the y-value. The point is at $(2, 5)$, so $f(2) = 5$.

13. $|-4| = 4 > |-1| = 1$, so Function B decreases faster.

14. Check the differences in y: $8 - 5 = 3$, $11 - 8 = 3$, $14 - 11 = 3$. Constant change means linear. The others have non-constant changes.

15. Slope: $\frac{-1-(-4)}{1-0} = 3$. y-intercept: -4 (from $x = 0$). So $y = 3x - 4$.

16. Standing still means distance doesn't change: horizontal line. Then walking at a steady pace means distance increases at a constant rate: line sloping upward.

17. Subtract 6 from x and add 3 to y: $(2 - 6, -1 + 3) = (-4, 2)$.

18. A 3×4 rectangle and a 2×6 rectangle both have area 12, but they are not congruent because their side lengths differ.

19. A dilation by factor k from the origin gives $(kx, ky) = (2 \times 3, 2 \times 4) = (6, 8)$.

20. Perimeter scales by the same factor. $45 \times \frac{2}{3} = 30$ cm.

21. Corresponding angles are in the same position (e.g., both upper-left) at each intersection of the transversal with the parallel lines.

22 $11^2 + 60^2 = 121 + 3600 = 3721 = 61^2$. Since $a^2 + b^2 = c^2$, it is a right triangle.

23 Both points are on the x-axis. Distance $= |7 - (-5)| = 12$.

24 $V = \pi r^2 h = \pi(49)(3) = 147\pi \ m^3$.

25 The independent (explanatory) variable goes on the x-axis; the dependent (response) variable goes on the y-axis.

26 The y-intercept is the y-value when $x = 0$. The line passes through $(0, 3)$, so $b = 3$.

27 A slope of 0 means no change in y as x changes; the line is horizontal.

28 Relative frequency is the ratio of a frequency to the total, often written as a fraction, decimal, or percentage.

29 Mean $= 4$. Every deviation $= |4 - 4| = 0$. MAD $= 0$. No spread.

30 An impossible event has probability 0 because it can never occur.

📋 Practice Test 7 — Answer Key

1 $\sqrt{2}$ (or $\sqrt{3}$)　　**2** C　　**3** 9 and 10　　**4** B　　**5** 6 cm　　**6** B　　**7** 317 times

8 C　　**9** B　　**10** C　　**11** 1　　**12** B　　**13** B　　**14** C　　**15** $y = 6x - 3$　　**16** C

17 C　　**18** C　　**19** B　　**20** $\frac{3}{2}$　　**21** B　　**22** B　　**23** 5　　**24** B

25 No. Association (correlation) does not prove causation.　　**26** B

Find more at
ViewMath.com/SC-Grade8

27 Predicted height at day 10: $y = 0.4(10) + 2 = 6$ cm. Slope: the plant grows 0.4 cm per day. y-intercept: the plant starts

28 B   **29** A   **30** C

> 💡 **Time to Learn!** 💡
>
> Review the explanations below, **especially for the questions you missed.**
>
> Understanding why each answer is correct builds stronger problem-solving skills.
>
> **Tip:** Circle any questions you got wrong, then read their explanation carefully.

📖 Practice Test 7 — Detailed Explanations

1 $\sqrt{2} \approx 1.414$ is between 1 and 2 and is irrational because 2 is not a perfect square. $\sqrt{3} \approx 1.732$ also works.

2 $\frac{36}{100} = 0.36$ (terminates). The correct conversion uses $99x = 36$, giving $\frac{36}{99} = \frac{4}{11}$. The student treated $0.\overline{36}$ as if it were 0.36.

3 $9^2 = 81$ and $10^2 = 100$. Since $81 < 83 < 100$, we have $9 < \sqrt{83} < 10$.

4 $2\sqrt{5} = \sqrt{4} \cdot \sqrt{5} = \sqrt{4 \times 5} = \sqrt{20}$. Alternatively, $2\sqrt{5} \approx 4.47$ while $\sqrt{10} \approx 3.16$. They are not equal.

5 Edge $= \sqrt[3]{216} = 6$ cm, since $6 \times 6 \times 6 = 216$.

6 Move the decimal 5 places left: $350{,}000 = 3.5 \times 10^5$. In scientific notation, a must satisfy $1 \leq a < 10$.

7 $\frac{1.9 \times 10^{27}}{6 \times 10^{24}} = \frac{1.9}{6} \times 10^3 \approx 0.317 \times 10^3 = 317$.

Find more at
ViewMath.com/SC-Grade8

8 In choice C, $\frac{9}{3} = 3$ but $\frac{15}{6} = 2.5$. The ratio is not constant, so it is not proportional.

9 $m = \frac{12-3}{4-1} = \frac{9}{3} = 3$.

10 $2l + 2w = 34$ and $l = w + 5$. Substitute: $2(w + 5) + 2w = 34$, so $4w + 10 = 34$, $4w = 24$, $w = 6$.

11 By the vertical line test, if a vertical line crosses a graph more than once, it is not a function. So a function's graph is crossed at most once.

12 $f(0) = -1$ and $f(2) = 7$. So $f(0) + f(2) = -1 + 7 = 6$.

13 Same slope means the lines go in the same direction; different y-intercepts mean they never cross. Parallel lines have equal slopes and different intercepts.

14 $6 - 3 = 3$, $11 - 6 = 5$, $18 - 11 = 7$. The y-differences are not the same, so the rate of change is not constant — nonlinear.

15 Substitute $m = 6$ and $b = -3$ into $y = mx + b$: $y = 6x - 3$.

16 A horizontal line means the output does not change. The height remains constant during that time period.

17 Moving every point the same distance in the same direction is a translation (slide).

18 In $\triangle ABC$: $\angle C = 180 - 55 - 80 = 45°$. Since $\angle C$ corresponds to $\angle F$ in congruent triangles, $\angle F = 45°$.

19 Multiply each coordinate by $\frac{1}{2}$: $(\frac{1}{2} \times 8, \frac{1}{2} \times (-4)) = (4, -2)$.

20 $\frac{9}{6} = \frac{3}{2}$, $\frac{12}{8} = \frac{3}{2}$, $\frac{15}{10} = \frac{3}{2}$. The scale factor is $\frac{3}{2}$.

Find more at
ViewMath.com/SC-Grade8

21 $180 - 54 - 54 = 72°$.

22 $6^2 + 8^2 = 36 + 64 = 100$, but $11^2 = 121$. Since $100 \neq 121$, it is not a right triangle.

23 $d = \sqrt{(7-3)^2 + (4-1)^2} = \sqrt{16 + 9} = \sqrt{25} = 5$.

24 $V = \frac{1}{3}\pi(4^2)(9) = \frac{1}{3}\pi(144) = 48\pi \ cm^3$.

25   Two variables may be associated due to a third variable or coincidence. Causation requires controlled experiments.

26   A straight line of best fit only works for data with a linear pattern. Curved data needs a different model.

27   Slope $= 0.4$ means 0.4 cm of growth per day. y-intercept $= 2$ means the initial height is 2 cm.

28   A two-way frequency table displays data for two categorical variables simultaneously, showing how frequently each combination occurs.

29   Lower MAD means less spread and more consistency. Class X (MAD $= 5$) is more consistent.

30   There are 13 hearts in 52 cards: $P = \frac{13}{52} = \frac{1}{4}$.

✅ Practice Test 8 — Answer Key

 C   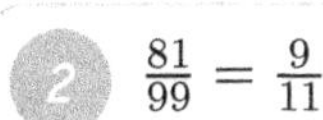 $\frac{81}{99} = \frac{9}{11}$  ≈ 6.7  D    -5  C    B    A    B

 16 and 20    B   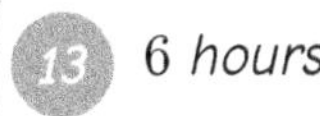 C    6 hours   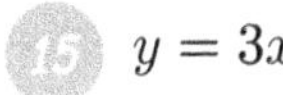 Nonlinear    $y = 3x$

Find more at
ViewMath.com/SC-Grade8

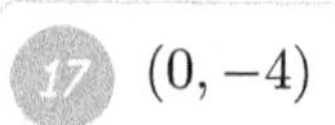 **16** The person saved less per month in months 4–6.

 **17** $(0, -4)$

18 C

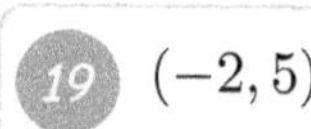 **19** $(-2, 5)$

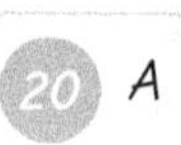 **20** A

21 $83°$

22 B

23 D

24 $288\pi \ in^3$

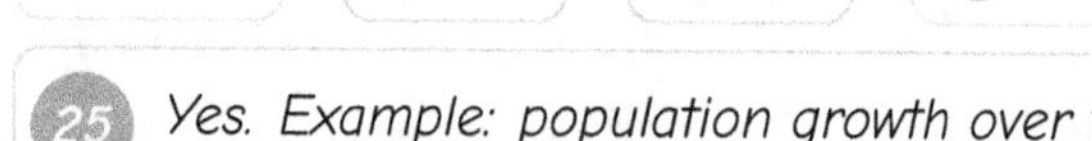 **25** Yes. Example: population growth over time may curve upward (exponential), showing a positive but nonlinear ass

 **26** C

 **27** 25 hours

28 B

 **29** $MAD = 6$

30 B

💡 Time to Learn! 💡

Review the explanations below, **especially for the questions you missed**.

Understanding why each answer is correct builds stronger problem-solving skills.

Tip: Circle any questions you got wrong, then read their explanation carefully.

📖 Practice Test 8 — Detailed Explanations

1 Every integer n can be written as $\frac{n}{1}$, making it rational. Not every square root is irrational (e.g., $\sqrt{4} = 2$).

2 Following the same pattern: $99x = 81$, so $x = \frac{81}{99}$. Simplify by dividing both by 9: $\frac{81}{99} = \frac{9}{11}$.

3 $6.7^2 = 44.89$ and $6.8^2 = 46.24$. Since 45 is very close to 44.89, $\sqrt{45} \approx 6.7$.

4 $\sqrt{3} \times \sqrt{12} = \sqrt{36} = 6$ and $\sqrt{6} \times \sqrt{6} = (\sqrt{6})^2 = 6$. But $\sqrt{3} + \sqrt{12} \approx 1.73 + 3.46 = 5.19 \neq 6$.

5 $(-5)^3 = -125$, so $\sqrt[3]{-125} = -5$.

6 $1 \times 10^2 = 100$ kg. From the chart, the lion's bar reaches close to the 10^2 mark, making its mass closest to 100 kg.

Find more at
ViewMath.com/SC-Grade8

7 $\frac{6.4\times10^{10}}{4\times10^{6}} = 1.6 \times 10^4 = 16,000$ *photos.*

8 *Machine A: 12 bottles/hour. Machine B: $\frac{50}{5} = 10$ bottles/hour. Machine A is faster.*

9 $\frac{50-68}{6} = \frac{-18}{6} = -3°F$ *per hour. The negative sign shows the temperature is decreasing.*

10 $x = y + 4$ *and* $x + y = 36$*. Substitute:* $(y+4)+y = 36$*,* $2y = 32$*,* $y = 16$*. Then* $x = 20$*.*

11 *Graph A is a circle and fails the vertical line test. Graph B is a parabola opening upward — every vertical line crosses it at most once, so it is a function.*

12 $f(-3) = -(-3) + 8 = 3 + 8 = 11$*.*

13 $8x + 20 = 3x + 50 \Rightarrow 5x = 30 \Rightarrow x = 6$*.*

14 *Check the y-differences:* $2 - 1 = 1$*,* $4 - 2 = 2$*,* $5 - 4 = 1$*,* $7 - 5 = 2$*. The differences are not all the same* $(1, 2, 1, 2)$*, so the rate of change is not constant. The function is nonlinear.*

15 *Slope:* $\frac{12-3}{4-1} = \frac{9}{3} = 3$*. Using* $(1,3)$*:* $3 = 3(1) + b \Rightarrow b = 0$*. Equation:* $y = 3x$*.*

16 *A less steep upward section means a smaller rate of change — less money was added per month compared to the first three months.*

17 *A 180° rotation uses* $(x, y) \rightarrow (-x, -y)$*. So* $(0, 4) \rightarrow (0, -4)$*.*

18 *Applying the 90° CCW rule* $(x, y) \rightarrow (-y, x)$*:* $R(0,0) \rightarrow (0,0)$*,* $S(4,0) \rightarrow (0,4)$*,* $T(0,3) \rightarrow (-3,0)$*. This matches* $R'S'T'$*.*

19 *Multiply each coordinate by* $\frac{1}{2}$*:* $(-4 \times \frac{1}{2}, 10 \times \frac{1}{2}) = (-2, 5)$*.*

Find more at
ViewMath.com/SC-Grade8

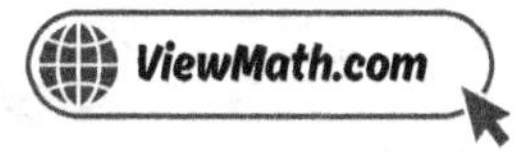

20 Dilation by $k = 2$ from the origin multiplies each coordinate by 2: $A(1,1) \rightarrow (2,2)$, $B(2,1) \rightarrow (4,2)$, $C(2,2) \rightarrow (4,4)$, $D(1,2) \rightarrow (2,4)$.

21 Co-interior angles are supplementary: $180 - 97 = 83°$.

22 The Pythagorean Theorem states that in a right triangle, $a^2 + b^2 = c^2$ where c is the hypotenuse.

23 $d = \sqrt{(4-1)^2 + (7-3)^2} = \sqrt{9 + 16} = \sqrt{25} = 5$.

24 $V = \frac{4}{3}\pi(216) = \frac{864\pi}{3} = 288\pi \ in^3$.

25 An association can be positive (both increase) and nonlinear (the trend is curved, not straight).

26 The line of best fit does not need to pass through any actual data point. It just needs to be close to most of them.

27 Set $y = 0$: $0 = -2x + 50$, $2x = 50$, $x = 25$.

28 $\frac{22}{40} = 0.55 = 55\%$.

29 Mean $= 60$. Deviations: $10, 5, 0, 5, 10$. $MAD = \frac{10+5+0+5+10}{5} = \frac{30}{5} = 6$.

30 $P(not\ event) = 1 - P(event) = 1 - \frac{3}{4} = \frac{1}{4}$.

✅ Practice Test 9 — Answer Key

 C    B    A    C    B    B    C    10 packages    A

Find more at
ViewMath.com/SC-Grade8

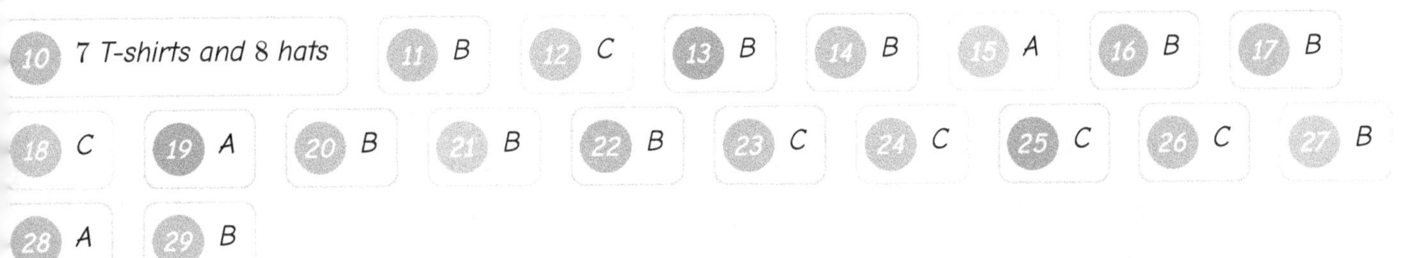

10 7 T-shirts and 8 hats   **11** B   **12** C   **13** B   **14** B   **15** A   **16** B   **17** B

18 C   **19** A   **20** B   **21** B   **22** B   **23** C   **24** C   **25** C   **26** C   **27** B

28 A   **29** B

30 Independent: flipping a coin twice (first flip doesn't affect second). Dependent: drawing cards without replacement (first

---

💡 Time to Learn! 💡

Review the explanations below, **especially for the questions you missed**.

Understanding why each answer is correct builds stronger problem-solving skills.

Tip: Circle any questions you got wrong, then read their explanation carefully.

📖 Practice Test 9 — Detailed Explanations

1 $1.41421356\ldots$ is the decimal expansion of $\sqrt{2}$, which is non-repeating and non-terminating, indicating an irrational number.

2 Let $x = 0.444\ldots$ Then $10x = 4.444\ldots$ Subtract: $9x = 4$, so $x = \frac{4}{9}$. Note that $\frac{4}{10} = 0.4$ (terminates), which is different from $0.\overline{4}$.

3 $6^2 = 36$ and $7^2 = 49$. Since $36 < 45 < 49$, we have $6 < \sqrt{45} < 7$. The student is correct.

4 $\sqrt{2} \approx 1.414$ and $\sqrt{3} \approx 1.732$, so their sum is approximately 3.146. $\sqrt{10} \approx 3.162$. Since $3.162 > 3.146$, $\sqrt{10}$ is slightly larger.

5 $7^2 = 49$ and $8^2 = 64$. Since $49 < 50 < 64$, we have $7 < \sqrt{50} < 8$.

Find more at
ViewMath.com/SC-Grade8

6 $8 \times 10^5 = 800{,}000$ and $3 \times 10^6 = 3{,}000{,}000$. The higher exponent wins: 3×10^6 is larger.

7 $7 \times 8 = 56$ and $10^{-3} \times 10^{-2} = 10^{-5}$. Then $56 \times 10^{-5} = 5.6 \times 10^{-4}$.

8 Service P: $4 \times 10 = 40$ packages. Service Q: $k = \frac{10}{2} = 5$, so $5 \times 10 = 50$ packages. Difference: $50 - 40 = 10$.

9 $m = \frac{7-(-2)}{3-0} = \frac{9}{3} = 3$.

10 $t + h = 15$ and $12t + 8h = 148$. From first: $h = 15 - t$. Substitute: $12t + 8(15 - t) = 148$, $4t + 120 = 148$, $4t = 28$, $t = 7$. Then $h = 8$.

11 Each student has exactly one student ID, so each input (name) maps to exactly one output (ID). That makes it a function.

12 $f(4) = 3(4) + 1 = 12 + 1 = 13$.

13 Hiker A's rate of change is 3 mph. Hiker B's slope is 4, meaning 4 mph. Since $4 > 3$, Hiker B walks faster.

14 A linear function has a constant rate of change (slope). Nonlinear functions have a rate of change that varies.

15 Slope $= \frac{10-2}{3-(-1)} = \frac{8}{4} = 2$.

16 A steeper upward slope means faster increase. The morning section is steeper than the afternoon section, so the temperature rose faster in the morning.

17 The $90°$ CCW rule is $(x, y) \to (-y, x)$. So $(2, 0) \to (0, 2)$.

18 Two rectangles can have the same perimeter but different dimensions (e.g., 2×6 and 3×5 both have perimeter 16), so they are not necessarily congruent.

Find more at
ViewMath.com/SC-Grade8

19 $90°$ CCW: $(x, y) \rightarrow (-y, x)$. So $(0, -5) \rightarrow (5, 0)$.

20 When a figure is dilated by factor k, its area is multiplied by k^2. So the new area is $20 \times 3^2 = 20 \times 9 = 180$ cm^2.

21 Alternate interior angles are equal: $4x = 80$, so $x = 20$.

22 $c^2 = 6^2 + 8^2 = 36 + 64 = 100$, so $c = 10$.

23 $d = \sqrt{1^2 + 1^2} = \sqrt{2} \approx 1.41$.

24 The volume of a cone is $V = \frac{1}{3}\pi r^2 h$, which is $\frac{1}{3}$ of a cylinder with the same base and height.

25 The dots go from lower-left to upper-right in a roughly straight pattern. This is a positive linear association.

26 $y = 1.5(4) + 2 = 6 + 2 = 8$.

27 Residual = actual − predicted = $8 - 6 = 2$.

28 Of 25 who did not study, 10 passed. $\frac{10}{25} = 0.40 = 40\%$.

29 Deviations: $|3 - 8| = 5, |5 - 8| = 3, |7 - 8| = 1, |9 - 8| = 1, |11 - 8| = 3, |13 - 8| = 5$. MAD $= \frac{5+3+1+1+3+5}{6} = \frac{18}{6} = 3$.

30 Independent events don't affect each other's probabilities. Dependent events do — the first outcome changes the conditions for the second.

✅ Practice Test 10 — Answer Key

1 B **2** D **3** B **4** D **5** C **6** Cell C, Cell A, Cell B **7** 3×10^6 **8** 99

9 B **10** B **11** 0, 3, 6 **12** B

13 F: rate of change = 2; G: rate of change = 3. Function G is greater. **14** B **15** B

16 Decreasing **17** C **18** $(3, -4)$ **19** A **20** C **21** B **22** B **23** 15 **24** B

25 A

26 Outliers can pull the line away from the overall trend, making it less representative of most data points.

27 B **28** B **29** B **30** A

💡 Time to Learn! 💡

Review the explanations below, **especially for the questions you missed.**

Understanding why each answer is correct builds stronger problem-solving skills.

Tip: Circle any questions you got wrong, then read their explanation carefully.

📖 Practice Test 10 — Detailed Explanations

1 5 is not a perfect square, so $\sqrt{5}$ is irrational and belongs in the Irrational region.

2 In Step 5, the student divided 2 by 10 instead of 9. From $9x = 2$, the correct answer is $x = \frac{2}{9}$.

Find more at
ViewMath.com/SC-Grade8

3 $7^2 = 49$ and $8^2 = 64$. Since $49 < 50 < 64$, we have $7 < \sqrt{50} < 8$.

4 $\pi \approx 3.1416$, $\sqrt{10} \approx 3.1623$, and 3.2 is exact. From least to greatest: $3.1416 < 3.1623 < 3.2$.

5 $\sqrt{2}$ cannot be expressed as a fraction of two integers, so it is irrational.

6   Cell A: 8×10^{-3} mm. Cell B: 5×10^{-2} mm. Cell C: 3×10^{-4} mm. From smallest to largest: $3 \times 10^{-4} < 8 \times 10^{-3} < 5 \times 10^{-2}$, so Cell C, Cell A, Cell B.

7 $\frac{3.6}{1.2} = 3$ and $10^{10-4} = 10^6$. Answer: 3×10^6.

8 $k = \frac{54}{6} = 9$. When $x = 11$: $y = 9 \times 11 = 99$.

9   A line that goes down from left to right has a negative slope.

10 $p + n = 8$ and $2p + 5n = 25$. From first: $p = 8 - n$. Substitute: $2(8 - n) + 5n = 25$, so $16 + 3n = 25$, $3n = 9$, $n = 3$.

11   The inputs are the values in the left column of the mapping diagram: 0, 3, and 6.

12   Set $6x - 4 = 0$. Add 4: $6x = 4$. Divide: $x = \frac{4}{6} = \frac{2}{3}$.

13   From the graph, F rises from $(0, 4)$ to $(2, 8)$: slope $= \frac{8-4}{2-0} = 2$. Function G has slope 3. Since $3 > 2$, G has the greater rate of change.

14   A linear function is written as $y = mx + b$, where m is the slope and b is the y-intercept.

15   Starting amount $b = 120$. Spending \$15/day means losing money, so $m = -15$. The function is $y = -15x + 120$.

Find more at
ViewMath.com/SC-Grade8

16   A graph that goes downward means the output (gas) is getting smaller over time.

17   Reflections are rigid transformations that preserve side lengths, angles, area, and perimeter.

18   Reflecting over the x-axis keeps x the same and negates y: $C(3,4) \to C'(3,-4)$.

19   First translate: $(1+3, 4+(-2)) = (4,2)$. Then reflect over the y-axis: $(4,2) \to (-4,2)$.

20   When $0 < k < 1$, the dilation is a reduction. $k = 0.25 = \frac{1}{4}$, so each length becomes $\frac{1}{4}$ of the original.

21   The three interior angles of any triangle always add to $180°$.

22 $a^2 = 17^2 - 8^2 = 289 - 64 = 225$, so $a = 15$. (Note: $\sqrt{225} = 15$, so B and D are the same value, but B is simplified.)

23 $d = \sqrt{9^2 + 12^2} = \sqrt{81 + 144} = \sqrt{225} = 15$.

24 $V = \frac{4}{3}(3.14)(125) = \frac{4}{3}(392.5) \approx 523.3 \ in^3$.

25   Clusters are groups of points that bunch together, while outliers are isolated far from the trend. Both can appear in the same scatter plot.

26   An outlier is an unusual point. Including it would distort the line and give a worse fit for the majority of the data.

27   The y-intercept ($b = 10$) is the predicted value when $x = 0$, meaning the predicted score with no study time.

28   Boys: 6 summer vs. 3 winter. Girls: 4 summer vs. 5 winter. Boys lean toward summer, girls lean toward winter.

Find more at
ViewMath.com/SC-Grade8

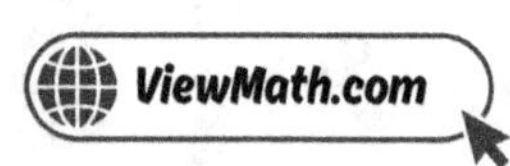

29. Multiplying all values by 3 multiplies the mean by 3 and each deviation by 3, so MAD is multiplied by 3.

30. With replacement: $P(red) = \frac{5}{8}$, $P(blue) = \frac{3}{8}$. $P = \frac{5}{8} \times \frac{3}{8} = \frac{15}{64}$.

Well done checking your answers!

Keep practicing to strengthen your skills.

Find more at
ViewMath.com/SC-Grade8

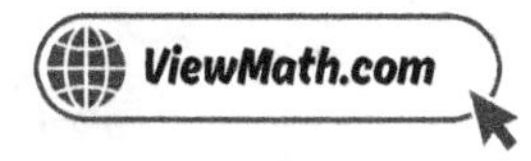

Author's Final Note

I hope you enjoyed this book as much as I enjoyed writing it. Whether you are a student working through the material, a parent supporting your child's learning, or a teacher guiding your class, I have tried to make this book as clear and engaging as possible. I hope I have succeeded. If you have any suggestions for improvement, please let me know. I would love to hear from you.

The accuracy of calculations is very important to me. We have done our best, but I also expect that I have made some minor errors. Constant improvement is the name of the game. If you find any errors, please let me know. I will fix them in the next edition.

For students: Your learning journey does not end here. I have written a series of books to help you learn math. Make sure you browse through them. I especially recommend workbooks and practice tests to help you prepare for your exams.

For parents: Thank you for investing in your child's education. I encourage you to explore the companion resources available online to help support your child outside the classroom.

For teachers: Thank you for the invaluable work you do every day. I hope this book serves as a useful resource in your classroom. Feel free to reach out if you have suggestions or would like to discuss how best to use this book with your students.

I also enjoy reading your reviews. If you have a moment, please leave a review on where you found this book. It will help others find this book. If you have any questions or comments, please feel free to contact me at drNazari@ViewMath.com.

And one last thing: Remember to use online resources for additional help. I recommend using the resources on `https://ViewMath.com` You can find video lessons, practice problems, and more. You can also use the online companion for this book to track your progress and access additional resources.

Wishing all students the best in their studies, parents every success in supporting their children, and teachers continued inspiration in their classrooms!

Dr. A. Nazari

Find more at
ViewMath.com/SC-Grade8

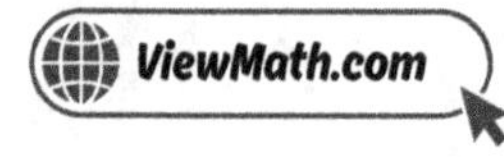